The Nine of Us

Painting by Ted Kennedy of the Kennedy home in Hyannis Port

The Nine of Us

Growing Up Kennedy

Jean Kennedy Smith

HARPER

An Imprint of HarperCollins*Publishers*

HarperCollins books may be purchased for educational, business, or sales promotional use. For information, please e-mail the Special Markets Department at SPsales@ harpercollins.com.

FIRST EDITION

Designed by Bonni Leon-Berman

Library of Congress Cataloging-in-Publication Data has been applied for.

ISBN: 978-0-06-244422-6

16 17 18 19 20   OV/RRD   10 9 8 7 6 5 4 3 2 1

This book is dedicated to

my outstanding and loving parents,

who were always there for us,

through the good times and the bad.

CONTENTS

PROLOGUE The White House by the Sea I

1 "No Whining in This House" 7

2 The Nine of Us I7

3 "No Irish Need Apply" 37

4 Closet Castaways 53

5 Faith, Values, and Hard Work 67

6 Grandma and Grandpa Fitzgerald 85

7 The Ocean in Our Veins 99

8 Our Jewel I09

9 On the Town with Dad I29

10 A Life Full of Lessons I45

11 Alone with Mother I63

12 The Dinner Discussion I75

13 Teddy I85

14 Daily Walks I93

15 Forever Changed 205

16 A Long Way from Bronxville 2I7

EPILOGUE And the Beat Goes On 237

ACKNOWLEDGMENTS 255

IMAGE CREDITS 259

The Nine of Us

The White House
by the Sea

The white house looked out over the sea. It was a sturdy and practical house, an overgrown Cape Cod cottage with white wooden shingles and black shutters, set back on a lawn that was worn in places from too many football games. A circular drive brought you up to the front steps, which ascended onto a long, wide porch. The beach waited just beyond the grass. A breakwall jutted out to the left to help calm the sometimes unruly seas.

A visitor climbing the steps in the late afternoon and pausing before the door might feel a moment of solitude, looking out at the water where gulls swooped down for their supper. But stepping inside, he would quickly realize that he was not alone at all. The white house was full. Full of activity, chatter, and laughter. Full of books on shelves and sports gear in closets. And especially full of children. Nine of us, to be exact.

We arrived each June in Hyannis Port with tremendous excitement. This was where, without fail and without question, the Kennedy children spent our summers. Like boys and girls all over the country, we hung up our satchels on the last day of school with satisfaction and relief, and we faced those carefree days with joy.

Just a few hours south of Boston by car, Hyannis Port seemed a world away from the rigors of our school years, which began for the older children in Brookline, Massachusetts, and then continued for all of us in Bronxville, New York. A small hamlet off the road from the larger town of Hyannis, it had only a simple post office to mark its place on the globe. Hyannis Port was so small that it probably had more boats in its harbor than people in its houses. Everyone knew one another.

Our house was a short bike ride out of town down Longwood Avenue and toward the sea. It started out as a cottage when Dad first purchased it in 1928. But it slowly grew as our family grew, its various additions rambling out from one side or another. In the summer, the house rarely got any rest, as we swung in and out of its doors. During the day, it served as our landmark on the shore, keeping us oriented while we swam or sailed on the water. At night our flashlights traced across the lawn where, like countless other children under the same moon in America, we scuttled among the bushes, playing hide-and-seek. On evenings when it was cool or the rain lashed outside, we gathered in the living room for fierce games of charades, acting out books, movies, or plays against the ticking clock. "It was *Great Expectations*!" I wailed after the time had run out without anyone picking up on what my wild hand gestures meant.

Hours later, after snacks and glasses of milk in the kitchen, we would move off to bed. Joe and Jack always had the bedrooms on the first floor, off the sunroom. For the rest of us it was a moveable feast, depending on who was home at the time. My older sister Kick and I often shared the room in the far corner of the house, at the end of the upstairs corridor. But if she happened to bring two or three friends home for the weekend, I would find myself in the room at the top of the stairs for a night or two. It was summer, and there was an easy flow to it all. With a few late-night whispers, we settled into bed, another day tucked in.

But what—another day so soon? The sun rose, the house stirred. Two feet hit the floor. Brisk steps down the hallway.

Joe, Jack, Rosemary, Kick, Eunice, Pat, Bobby, and me.
Teddy was not yet born (1931)

Coffee on the stove. Dad, ready and alert in his riding clothes, with Teddy fast on his heels, filed out the front door, headed for their morning horseback ride through the cranberry bogs in nearby Osterville. Mother rose as well, said her morning prayers, and dutifully prepared for the short drive to St. Francis Xavier Church in Hyannis for daily Mass. Eunice, Bobby, or I sometimes awakened to go along with her.

Or else we didn't, and instead stayed burrowed in our pillows, along with everyone else, for just a few more minutes of precious sleep. Then we stretched our arms up and out, ready for whatever the day might bring, under the care and in the fold of the big white house.

"No Whining in This House"

*There is no other success for a father and a mother
except to feel that they have made some contribution
to the development of their children.*

—JOSEPH P. KENNEDY

*P*eople have asked me for years about my family, about why
we are so close, about how it happened that we all became involved
in politics and interested in the issues of the world. They are naturally curious about life with my sisters and brothers, who all grew up
to be well known, but who, in the beginning, were simply children.

Like the people who have asked me these questions, it is interesting for me to reflect on those earlier years and to think about
what, if anything, made them beyond the ordinary. For in so many
ways, our life seemed unexceptional most of the time. Even though

I was there through it all, it is hard for me to fully comprehend that I was growing up with brothers who would eventually occupy the highest offices of our nation, including president of the United States. At the time, they were simply my playmates. They were the source of my amusement and the objects of my admiration.

*J*ust as in any family, we had our happy days and our moody ones. We teased one another mercilessly. We argued over who deserved the last piece of cake. We planned exploits into forbidden territory, climbing too high in trees or onto the garage roof. And we laughed. I can say without reservation that I do not remember a day in our childhood without laughter.

Certainly a distinct characteristic of our family was its size. Growing up in a big family of nine children is a less common experience today than it was in those days, and it certainly left its mark on each of us. A large family can be a challenge, particularly for parents whose attentions are inevitably divided and whose patience must be stretched very thin at times. But any challenge of a big family is eclipsed by the tremendous fun. A child in a big family constantly feels surrounded and supported. For me, there was always someone to play with or someone to talk to just around the corner, out on the porch, or in the next bedroom. I never felt alone.

Our childhood played out during a different time in America, when children found their amusement conspiring in closets, playing ball in the front yard, or lost in the pages of an adventure novel. This was the heyday of radio. Televisions, computers, and video games had not been invented yet, and it was beyond our imagination that anything like that would ever exist. We did not automatically learn what was going on in the next town, state, or country by turning on a screen. Instead we waited for the newspaper to arrive on the porch each morning. Telephones were firmly attached to the wall by a thick black cord, and they were treated with respect. Time on the telephone equaled money. When it rang, you usually knew it was signaling exciting news or, hopefully not, a serious emergency. Either way, you never talked on the telephone for very long.

To get around town or to school, we walked or jumped on our bicycles. For longer trips, we rode in cars. And for even longer trips, we boarded great locomotive trains or vast steamer ships. Air travel was almost unheard of, and an enormous luxury. And no matter how we got where we were going, we always dressed up. Gloves and hats and jackets were typical attire, not just our Sunday best.

In the early decades of the twentieth century, when my brothers and sisters and I were born, the United States was still taking its shape as a nation. It was not yet the global powerhouse it would soon become. We grew up in those disquieting years between World Wars I and II, when America was in constant motion, an ambitious young nation wresting its way out of the Depression. We were aware that trouble was swelling overseas, although for us little ones in the family, it was more an abstract concept than a reality.

Central to our lives and to the people we would become were our parents, Rose and Joe Kennedy. They influenced the nine of us so profoundly, yet so subtly, that I hardly understood the impact myself until I looked back decades later through mature eyes. Mother and Dad both descended from Irish immigrants, a fact that profoundly affected their outlook on life and the choices they made for their family. They were very conscious of the tremendous oppression their ancestors had overcome and were extremely grateful to be Americans. Like other young people of their generation who were the grandsons and granddaughters of immigrants, Mother and Dad felt a duty to give back to the country that had embraced their family, and to contribute to its continued growth.

"To whom much has been given, much is expected."

Mother repeated those words to us often, quoting the Gospel of St. Luke. She was not admonishing us. She was challenging us. Mother and Dad felt it was important for us to know how lucky

Mother and Dad (Hyannis Port)

we were. "On some days, your ancestors had no idea where their next meal was coming from," Mother would remind us. In contrast, we had a roof over our heads and we had hot food on our table every night. Mother and Dad let us know that we had an obligation to give back.

"To whom much is given . . ."

All of us understood from the earliest age that we were required to use our talents and gifts for the good of others and of our country. There was no other option. For the gift of being in this world, we had a responsibility to it. This is what motivated Mother and Dad in their daily lives, and it was what shaped how they raised us. Theirs was a constant reminder: take nothing for granted, work hard, and be grateful.

Yet despite the seriousness of this obligation to them, somehow Mother and Dad made it all seem fun and light and interesting. They did not make unrealistic or unkind demands or hold our noses to the grindstone. I do not remember them once raising their voices to us—ever. Rather, they compelled us to be our best selves.

"I definitely know you have all the goods and you will go a long way," Dad wrote to my brother Jack in his typical, encouraging tone.

Mother and Dad's approach was steady, yet firm. And they were purposeful. As I look back, it is clear how intentional my parents were in every decision they made. They did not rely on chance, but instead set very clear rules through their example.

Complaining was strictly forbidden. We were not allowed to sit around moaning because we could not go to the movies or had received a poor mark in geometry class. If Teddy got more cookies, if Pat borrowed my bicycle without asking, if things seemed unjust or unfair, we learned early on that the response was not to grumble or cry. Dad's voice would clamp down on our ears: "Fix it. There's no whining in this house." He could not abide us feeling sorry for ourselves. Life was far too good for us to whine about small things. It was selfish, and on top of that, it was boring. Why should everyone else have to suffer through your complaints about homework?

This belief, this mantra, was so imbedded in our family ethos that even as we grew up and faced larger problems, even if we were a thousand miles away, when a whine escaped our lips, we could hear him: "There's no whining in this house."

Mother and Dad felt that in order for us to understand the people and events around us, we needed a strong grasp of the people and events that came before. They made sure that our Grandpa and Grandma were fixtures in our childhood so that we had an appreciation for life outside our generation and an understanding of the odds that they and their parents faced. At Thanksgiving each year, Mother and Dad told us the story of the brave Pilgrims of Plymouth, and of the equally brave Native Americans who overcame what must have been great fear and uncertainty to help the Pilgrims in their moment of need.

Our trips to Boston and elsewhere in the state were history lessons in disguise as Mother spun fascinating accounts of the founding of our nation: the Old North Church, the Battle of Bunker Hill, Lexington and Concord. A born teacher, Mother peppered us with questions: How long do you think it took to ride to Concord, children? Who do you think fired the first shot? Why do you think Thomas Jefferson felt the pursuit of happiness was a fundamental right?

In the mornings over breakfast, Mother served up the news of the day along with our cereal and toast. The image of her still makes me laugh, arriving at the table with newspaper articles that she found interesting pinned to her dress. Jack's boyhood friends later marveled that he was the only fourteen-year-old they knew who had the *New York Times* delivered to him at boarding school each day. Mother had made current events so interesting at home that we would not think of missing out when we were away at school.

Mother and Dad taught us how to take care of one another without telling us to. They taught us how to love one another without forcing us to. Their example was most apparent when it came to our sister Rosemary. Rosemary was born with intel-lectual disabilities that we did not completely comprehend until years later. As a young child, I simply understood that she had greater trouble doing certain things than the rest of us did. But I also understood from my parents, as we all did, that Rosemary was an integral part of our lives. Mother and Dad did not have to tell us directly. We just knew how to act by how they acted. When doctors suggested placing Rosemary in an institution, Dad's reply was immediate: "What can they do in an institution that we can't do better for her at home, here with her family?" So she stayed with us. It would never have occurred to us to leave Rosemary behind or leave her out—because she was fun and we loved her.

When I was grown, with a family of my own, I asked Mother what job she might have pursued had she been born at a time when women had more opportunity in the workforce.

"Mother, I could see you being a college president," I said, thinking of her thirst for teaching and learning.

"Oh no, Jean, dear," she replied, shaking her head. "I have no regrets whatsoever. A college president is a wonderful profession. But for me, raising a child is the most challenging and rewarding profession of all."

How grateful I am—how grateful we all were—that Mother and Dad took that singular point of view. They knew how to cure our hurts, bind our wounds, listen to our woes, and help us enjoy life. We were lucky children indeed.

Seated from left: Pat, Bobby, Mother, Jack, and Dad, with Teddy on his lap
Standing from left: Joe, Kick, Rosemary, Eunice, and me (Hyannis Port, 1930s)

2

The Nine of Us

*In our family, in sickness and in health, we were
all involved with one another, all in the same life, a
continuum, a seamless fabric, a flow of time.*

—ROSE FITZGERALD KENNEDY

At any time, in any generation, nine children is a lot.
Today, it is rarely heard of. Yet it seemed that, from the start,
Mother and Dad were destined to have a gaggle of children. They
loved having children around them, and they worked well together as parents and partners. We would not have been complete
if they had stopped at two or four or even six. Nine of us we had
to be. And each of us had our singular spot in line.

Mother kept track of all our vital statistics on index cards that became an absolute necessity as our number began to grow. She said she got the idea while walking past a stationery store one day, early in her marriage. Her head must have been full of all the facts she had to remember for each of her children back at home. She purchased her supplies and, from that point on, when a new baby arrived, a new card went into the box, marked first with simply birth date, weight, and length. But soon our distinctive lists of statistics began to grow, updated each week without fail: date of baptism, names of godparents, vaccinations, dental appointments, First Holy Communion, and any illnesses or broken bones that might befall us.

The inaugural card in Mother's box belonged to Joe— handsome, bold, brilliant Joe, who, first in the world, became first at everything. Named after Dad, he was dogged, loyal, and keenly intuitive like his namesake. He had an insatiable love of life and threw his heart and soul into everything he touched, bringing home both a broken nose and the championship trophy in foot- ball, landing on the dean's list at Harvard and heading up every activity he could think to join, from Debate Club to Class Day Committee. He was exceptionally intelligent, gallant on a horse, and strikingly handsome. And he made everything seem so effort- less and interesting. We all wanted to follow in his footsteps. Joe was a magnet, for his friends and for us.

Joe (1936)

Joe seemed to know that we were looking up to him from our successive spots in the family, and he took very seriously the responsibility that Mother and Dad placed on him to lead us in the right direction. His college friend Timothy J. Reardon once recalled Joe telling him, "You know, T.J., I'm the oldest of my family and I've got to be the example for a lot of brothers and sisters."

I always liked to think that Joe held a special place in his mind and heart for me because I was his godchild. Right after I was born, Joe asked Mother and Dad for the job. At the age of thirteen, he felt he was old enough to be my godfather, and they readily agreed. Mother told Joe that being a godfather was a serious commitment. He needed to protect me and watch out for me. And so he did. When he was away at boarding school, Joe wrote me regular letters inquiring about my marks and whether my brothers and sisters were being nice to me. And when he came home, he made it his mission to teach me how to ride a bicycle. Down the hill he sent me, high on my mechanical steed. But he failed to tell me how to stop. So smack into a tree I went. Poor Joe. I probably was not the easiest godchild. But ride a bike I finally did.

Not two years after Joe's birth, the number two child arrived: Jack. Where Joe was solid and broad, Jack was slim and wiry, but equally lively and tough. He was funny and original, charting his own path regardless of what others thought. When he was a little boy, Jack had what mother described as an "elfin quality," and his photographs prove her right. He was sick with the ailments that plagued a lot of children at that time. Whooping cough, measles, chicken pox, and the dreaded scarlet fever all found Jack and sent him to bed.

Jack at the Cape (circa 1940)

It was under his covers recovering that Jack developed his vo-racious and lifelong love of reading. There were no electronics to occupy him, so books were his sole form of entertainment. The characters Jack got to know on those pages were often hapless charmers full of mischief. He loved Reddy Fox, Peter Pan, and Billy Whiskers. One of his favorite books was *The Story of a Bad Boy*, in which a young scamp named Tom Bailey pulls off a series of adventures with his motley gang of friends. As he grew older, Jack became enthralled with history and legend: *King Arthur and His Knights, Lays of Ancient Rome, Treasure Island, The Arabian Nights, The Writing and Speeches of Daniel Webster*—I am convinced that Jack's singular humor, his inquisitiveness, and his love of history were set in stone during those long hours beneath his blankets with the characters from those books.

Jack took books everywhere he went. I would often find him on the porch or in the corner of the living room deep within the pages of his latest volume. There was no chance of getting his attention no matter how hard I tried. But then he would reach the end of a chapter and—snap!—the cover would close and Jack was back on the move. Off for a swim. Off to play tennis. Off anywhere with Joe, his constant companion.

Joe and Jack's friendship was legendary, marked by escapades and practical jokes, as well as occasional spikes of temper. Joe's fuse was particularly short. Jack once famously borrowed Joe's swimming trunks, causing our eldest brother to chase him down with a fury, straight out to the breakwall in front of the house. They undoubtedly would have wrestled all the way into the sea but for Dad's close friend and secretary, Eddie Moore, who was sitting on the beach and quickly jumped up to pull the two of them apart.

Rosemary (1938)

Yet my abiding memory of Joe and Jack is at the dinner table each night, where they fed off each other's energy, debating the critical issues of the day with Dad. Who would win the election? Would there be war in Europe? Should the United States intervene? We younger children sat at a smaller table nearby, watching and listening, wanting to know what they thought and hanging on each gripping word.

Soon to follow after Jack was Rosemary, named for our mother, Rose. Gentle and kind, Rosemary—or Rosie, as Dad called her—had a beautiful Irish face and smile. When I was small, she was one of the older group, one of the sisters who would shepherd me from place to place. Only as I grew did I begin to realize that Rosemary had challenges that the rest of us did not have. The only words the doctors had to describe her condition were "mental retardation." She had a hard time keeping up in class and finishing her work. It naturally frustrated Rosemary, and although Mother and Dad searched for a school to help her, there were few, if any, at that time that specialized in teaching children with disabilities. So Rosemary struggled on. In the summers, though, she was just one of the gang.

Rosemary laughed with the rest of us and swam with the best of us. She was at her happiest when we were all together. Rosemary was crazy about Dad, and he was equally crazy about her, solicitous and gentle, always keeping an eye out for her and waiting behind for her to catch up. Joe and Jack escorted Rosemary on summer nights to the Wianno Club nearby in Osterville, where they took turns swinging her around the dance floor. And I spent steamy days hitting balls back and forth with her on the tennis court. Rosemary was a formidable, though always sweet, opponent. She worked very hard and tried very hard, and we always had a good time.

Kathleen was next. Mother and Dad nicknamed her Kick, a perfect name for her, since she kicked off her covers and jumped into the world with vigor every morning. There was no higher

Kick (1938)

spirit than Kick. I do not remember her as anything but out-going and exhuberant. People of all ages were drawn to her. Like Rosemary, Kick was close with Dad, who adored her lively air. And she seemed perpetually positioned between Joe and Jack, on their arms and off to one party or another. They would travel in a pack: the boys escorting Kick's girlfriends and Kick being escorted by their pals. It always seemed like the perfect arrangement.

I felt especially lucky in life because Kick and I shared a birth-day. The youngest daughter, I was born on February 20, 1928, the day that Kick turned eight. "I like Jean very much," she wrote to Dad shortly after my arrival. And she made that abundantly clear to me at every turn. Each year, we had a joint birthday party with cake and ice cream and friends of all ages. And each year, without fail, Kick would take me aside and give me a special pres-ent that she had chosen just for me. Though much older and more glamorous than I, with so many friends and beaux in her swirl, Kick never forgot our connection.

Sporty Eunice was number five. Was there nothing she didn't try? Football, tennis, sailing—Eunice was game for anything, and she was in it to win. Our brothers, one more competitive than the next, had tremendous admiration for Eunice's sporting ability. I envied how, when preparing for a sailing regatta, they went look-ing for Eunice to join the crew, sometimes even offering her the captain's spot. One year, she traveled to Pennsylvania and Ohio to play in tennis tournaments, reporting in her letters home that her victories in the matches were broadcast in "big black type" in the newspaper. "Wait 'til you see that Eunice gal catch passes," Joe told a friend as they headed down the steps to the lawn for a game of touch football.

In the whirl of her day, Eunice had no time for nonsense or

Eunice (1935)

small talk. She was direct, matter-of-fact, and a force. She had so much energy you could feel her coming minutes before she arrived. And always close by her side was Rosemary. Eunice had a special love for Rosemary. Whatever she did, Rosemary did. Look out the window and there they were, side by side, younger sister making sure older sister always was in the mix. This certainly endeared

Pat (1935)

Eunice to Mother, who naturally worried about Rosemary and was happy she was safe in her sister's care.

Patricia—"Pat"—glided into life at number six. Instantly enticing, she was tall and stunning, but not in the flashy way of a film star or ingénue. Pat was understated. She did not need a gown to make an impression. She was a knockout in a sweatshirt and dungarees. Everyone took notice of Pat because of her looks, but also because of her joie de vivre. If you met Pat, you loved Pat. She had impeccable taste and a great love of the dramatic arts, no matter what the medium—theater, film, music. When I was young, I often found her captivated by the latest edition of *Variety*, the show business magazine Dad had helped start. He also invested in plays and would bring the scripts home for Pat to read and critique.

Pat was always an attentive older sister to me and such fun to be around. One time, when Mother and Dad were away, Pat and I were left in the house alone with Margaret Ambrose, our amazing and generous cook, who took the greatest pleasure in feeding us. Each night we gleefully devoured what in retrospect was a very odd, yet irresistibly delicious, dish: fried chicken and waffles drenched in pure maple syrup. We followed that up with mile-high chocolate cake covered with chocolate sauce and ice cream. When our parents returned home, they were shocked and dismayed to find two very plump daughters. But Pat and I had never been happier.

Next up, at lucky number seven, was a character named Bob. Adventure, sport, and mighty quests filled his every hour. He was great fun, but his days were never futile nor frivolous. Bobby seemed to always have a mission behind everything he did. He made friends effortlessly yet intentionally. And once a person was

Bobby (1935)

his friend, Bobby was loyal for life. He had compassion for everyone. His friendships extended beyond the human variety to pet rabbits, dogs, and a pig. Bobby rescued chipmunks, birds, frogs, and any other creature he might happen upon on the sidewalk or in the woods. It was as if he were born with a purpose that he steadfastly pursued: to impact the world and the things around him, no matter how small.

Trouble seemed to follow Bobby more than the rest of us. Our nurse, Kikoo, would groan when Bobby got tar in his eye or fell off the roof. He seemed always to have a bruise or a broken bone from either an accident or a tussle on the playing field. For Bobby was fiercely competitive, and that competitiveness underpinned everything he took an interest in, from baseball to current events. I can still hear his familiar cry when we were united in a game of touch football against a team of friends: "Let's go Kennedys!" Bobby also strove to be as near as possible to Joe and Jack every chance he got, and to be respected by them. They were many years older than he, and engaged in important things. At dinner-time, at the kids' table off to the side, Bobby strained his ear in their direction and longed to be their equal.

I sneaked in at number eight—Jean, the sidekick, the partner in crime to the antics of everyone around me. Mother claimed I was the only child ever to directly tell her no. I have never liked that distinction but accept it as truth because Mother would not have gotten it wrong. Perhaps I was being egged on by my number one playmate, my lieutenant in life, the barrel of laughs that came after me, number nine . . .

Teddy. Adored by all. Happy, carefree, playful, and loveable— Teddy was born with huge cheeks that called out to be squeezed. By the time Teddy was born, Dad had achieved his fair share of

Mother holding me, with Bobby (1928)

success in business and politics, so even as a tiny boy, Teddy was spending time with some of the most prominent people in world history: senators, congressmen, film stars, the president, the Pope. He even met Babe Ruth at Yankee Stadium. And because he was Teddy, a smaller version of our charismatic Grandpa Fitzgerald, he was completely at ease in their company.

The rest of us spent our days in stitches thanks to Teddy. One summer, Mother enrolled Teddy and me in boxing lessons at the Bath and Tennis Club in Palm Beach, Florida, where we spent

Teddy and Bobby (1934)

winter vacations. We danced around the ring across from each other, leather gloves held up to our faces, neither willing to punch the other, until we collapsed on the mat in giggles. Our burly boxing coach was not amused.

Teddy wanted to be with the older boys every chance he got, and they were happy to let him tag along. They introduced him to the sea, teaching him to swim and, to his life's delight, sail. From early on, Teddy seemed always happiest on the sea. He spent most of his summers off the coast of Hyannis Port, on a boat, his smile wide, his chatter constant, and his laughter roaring like the waves.

Those were the nine of us. And at the helm were Mother and Dad. Mother, a petite woman with an indomitable nature and a sure understanding of what we needed in life. Dad, a towering presence, always with the right answer for our worries and a tender place for our growing spirits. They were our leaders, teach-ers, and champions. They made everything possible. They made everything clear. Our story is theirs. Falling in line, matching their step, into life we marched.

RIGHT Back from left: Eunice, Joe, Jack holding me, and Rosemary
Front from left: Bobby and Pat (Hyannis Port, 1934)

3

"No Irish Need Apply"

*They arrived with scarcely more than determination
and faith . . . they worked hard and raised their
children in the love and care of God.*

—ROSE FITZGERALD KENNEDY,
ON HER IRISH ANCESTORS

$\mathcal{M}$other and Dad did not start out planning to have a
house by the ocean. They were both born in the city of Boston,
and in Boston they were bred.

Their grandparents were among the thousands of immigrants
who landed in the bustling port city on the Charles River in the
late, tumultuous years of the nineteenth century. They left for
the only reason anyone would ever leave Ireland: they were starv-
ing, and desperate for work. Born in rural Irish counties, these

Grandpa John "Honey Fitz" Fitzgerald (Boston, 1906)

young men and women, carrying the surnames Hannon, Fitzgerald, Hickey, and Kennedy, were on the cusp of adulthood when a blight known as the Irish Famine descended on their country. This period of mass starvation, disease, and emigration is considered one of the greatest tragedies in European history, one that forever changed the island of Ireland and the future of our family.

More than one million Irish people died. Another million were forced to emigrate, boarding so-called coffin ships for far-off destinations such as America and Australia. They left behind fruitless fields, stoic fathers, and distraught mothers who tucked their last pence and their last hopes into the pockets of the sons and daughters embarking on those ships. My great-grandparents were among them. They hailed from counties Wexford and Limerick. They all shared the same desire to make a better life for themselves in America.

Once they arrived, these young Irish immigrants flooded the ports and streets of Boston. Like millions who came before and after them, from towns and villages around the world, they sought jobs, food, and solace in a country built on the principle that everyone is welcome. Yet like millions who came before and after them, they quickly found that the promise of America was tarnished by anger, distrust, and fear. Members of the Boston establishment, descendants of immigrants themselves, were not welcoming of these new arrivals.

Some Boston shopkeepers made their disapproval of the Irish abundantly clear in the signs they hung in their shop windows: "No Irish Need Apply." "No Irish Need Apply" greeted my two grandfathers when, as young men eager to excel, they went out on the streets of Boston looking for work. "No Irish Need Apply" still hung in the windows of shops when my mother was a young girl. As a child, I often imagined Mother walking along the city streets with Grandpa, her tiny hand holding fast to his giant mitt. In my imagination, just as they approached an unwelcoming shop, Grandpa would steer Mother across the road to the opposite sidewalk, saving her the humiliation of seeing the sign hanging in the window, directed at them.

Despite this outward hostility, just one generation after their parents stepped on Boston soil, my Kennedy and Fitzgerald grandfathers had earned places of prominence in that tradition-ally Brahmin town. Dad's father, Patrick Joseph, or P.J., started out working on the city's docks and then went into business on his own when he opened a local pub. Mother's father, John "Honey Fitz" Fitzgerald, rose to be a singular force in Boston politics. He served in Congress and, to the dismay of some of the city's establishment, was elected the city's first Irish Catholic mayor.

As a young woman coming of age in Boston, and the Mayor's daughter, Mother was the toast of the town. Yet she still felt something was amiss. As she started to make friends, she realized that the clubs that welcomed other ladies of the town did not open their doors for her, despite her connections to City Hall. At the turn of the century, there was still an unspoken rule that Irish

P. J. Kennedy

Catholics were not allowed in the social and business circles of Boston. Ever a pragmatist, Mother was not angry or hurt by these realities. She decided that if you were not accepted into one club, then you should just start one of your own.

So in 1910, Mother initiated her own ladies' club, the Ace of Clubs, where young, unmarried Catholic women could meet, have

Dad, early in his career (1926)

tea, and engage with guest speakers who addressed the matters of the day. As the club's records show, one week the ladies welcomed *Boston Herald* columnist E. E. Whiting, who shared his "wide knowledge of political situations." Two weeks later brought a "Mrs. Crawford," who spoke on "European situations, including the League of Nations." The club was an immediate success, and soon membership in it became highly prized; the Ace of Clubs remained open for one hundred years to follow.

In 1914, Mother would marry Dad, who was equally independent and strong-willed and shared her disregard for social barriers and her determination to rise above them. The young couple bought their first home at 83 Beals Street in Brookline, about a twenty-five minute trolley ride from the center of Boston. It was an old clapboard house with nine rooms. There Joe, Jack, Rosemary, and Kick were born. Once the quarters grew too small, the family moved to 131 Naples Road, a few blocks away, in time to welcome Eunice, Pat, and Bobby.

But it nagged at Dad that by growing up in Boston, his children might face the same obstacles and high-handedness from the establishment that he and Mother had encountered. "Boston was no place to bring up Irish Catholic children," Dad once told a newspaper reporter. He wanted us to have a carefree childhood where we would feel unencumbered and accepted. So Dad went in search of a second home, a summer home, where we would feel accepted no matter what our roots, no matter where we came from. Dad traveled south all the way to Cape Cod until he reached Hyannis Port. There he found the white house by the sea.

"If you want your children to come home, buy a house by the sea," Dad would later say. He was speaking from experience. Wherever we lived as a family, wherever we ventured, down the street to the village or across the ocean to Europe, the nine of us always came back to the sea, which was Hyannis Port.

There were other notable homes in our childhood. For me,

Pat and Teddy (Bronxville, 1934)

second only to Hyannis Port was the house in Bronxville. The year before I was born, Dad's work demanded that he spend a great deal of his time in New York City, away from Mother and his sons and daughters. This was not an ideal situation for a man who adored his family, which, by then, had seven children under the age of eleven. Dad could not stand being away for long, and Mother needed his input and help. So in 1927 they decided to

Teddy, Bobby, and me dressed as pirates on Halloween (Bronxville, 1934)

move the family closer to Manhattan. After some looking around, they settled on a small village just north of the city, in Westchester County, called Bronxville.

The home they found on Pondfield Road was large enough to hold the entire family. Anchored by soaring pillars out front, the Bronxville house had six acres of sweeping, wooded grounds that were perfect for exploring on fall afternoons, with a steep hill for sledding when the snow eventually came. We had enough bedrooms for everyone, with each one opening onto a wide second-floor balcony.

Bronxville is where our family spent the fall and winter months for more than a dozen years. It is also the house where Teddy and I first lived as infants, though we were born far away. As much as Mother loved New York, she was still a New Englander through and through, and she entrusted the delivery of her babies to only one obstetrician, Dr. Frederick Good of Boston. So when the time came for me to arrive into the world, in 1928, and for Teddy to arrive four years later, Mother left New York and traveled the two hundred miles to St. Margaret's Hospital in Boston to deliver us there. Then, after the specified period of recuperation, back to Bronxville we came.

Bronxville was the home where I learned to walk and from which I first went off to school. It was the scene of many family celebrations, including that most miraculous of holidays—Christmas! Mother once described Christmas as "the greatest event in our house," and she and Dad made it so. The entire family came home to Bronxville from our various boarding schools to a house full of activity. Dad would choose the largest Christmas tree he could find and set it up in the sunroom. It was then our job to drape it with tinsel and our colorful collection of ornaments. Construction paper and crayon creations we made for Mother hung side by side with the delicate, sparkling glass bulbs that she had received as gifts. Surreptitious gift wrapping took place in every corner of the house. "It's the prettiest tree yet!" Dad declared each year, as the presents began to pile up beneath it.

No matter how young we were, we liked giving gifts to one another. We might draw a picture and place it in a frame we had made from cardboard. Or we would head off to the five-and-dime store with Mother to find a little book, a set of pencils, or a toy car that Teddy or Bobby or Pat might like. Mother emphasized that it was the sentiment, not the size of the gift or price that mattered. She also made sure we understood that we were to be grateful for everything we received.

On Christmas Eve, Mother took her place at the piano in Bronxville and accompanied us through our most beloved carols: "Deck the Halls," "Silent Night," "Away in the Manger." At the top of our lungs, off-key and on, we sang, over and over again, verse after verse. Then, just before the smallest of us headed up to bed, she would gather us around the crèche and tell us the Christmas story. "This is the reason we are celebrating tonight," she explained. Her gentle voice spoke of the poor couple setting off toward Bethlehem more than a thousand years before, finding no place to sleep but a manger. With words, she painted a picture of the shepherds, late at night, looking up to see a blazing, yet welcoming, star that guided them across the fields to the humble stable. There, with Mother at the crèche, we felt warm, loved, and happy.

Christmas morning started with early Mass for all. We crammed into the pews in our wool coats. I wedged in between Eunice and Kick and itched for Mass to end so I could race back to the house and see what Santa Claus had delivered. The prayers seemed so much longer on Christmas morning. The priest moved so much more slowly. Didn't he have gifts to open, too?

Santa brought to us what he was delivering to boys and girls all across America. Each year we received just one special present: a doll, a game, a set of roller skates. It was impossible to contain Bobby's glee when he finally got the electric train he had been hoping for. It wound its way under the tree and out into the hall. For years to follow, he and Teddy would spend hours playing with that magnificent train upstairs in the crowded attic.

If we were lucky enough to have snow on Christmas, we were soon out the door and into the garage to retrieve our sleds. Soaping up the runners for maximum speed, we careened down the hill outside, again and again and again.

Another big Bronxville moment was the visit of Cardinal Eugenio Pacelli, who came one afternoon for tea. For a devout Catholic family like ours at that time in American history, this was a momentous occasion. As secretary of state for the Vatican, the cardinal had been touring the United States and was visiting President Franklin D. Roosevelt and other dignitaries at the president's home, Hyde Park, on the Hudson River. Much to my parents' delight, the American Cardinal Francis Spellman, a friend of our father's, had arranged for Cardinal Pacelli to stop for a visit at our house on his way back into New York City.

We all stood in a line in the front hallway at the Bronxville house to greet His Eminence, curtseying and bowing and then kissing his ring as Mother had instructed us. Cardinal Pacelli wore long red robes and the trademark magenta skullcap. We were nervous, but he quickly put us at ease with his gentle nature.

The cardinal moved into the living room, where he sat on our sofa. I watched with envy as he took four-year-old Teddy on his knee. We, of course, had no idea at the time that in just three years, Cardinal Pacelli would be elected by the College of Cardinals as Pope Pius XII. By that time, our father would be the U.S. ambassador to the Court of St. James and we would be living in London. We traveled to the Vatican in Rome, where, again to my envy, Teddy received his First Holy Communion from the newly elected Pope.

Mother had a plaque made and mounted on the back of the living room sofa to mark the day the future Pope sat there. Once Mother and Dad sold the Bronxville house, they moved the sofa to Hyannis Port, where it held a place of honor in that living room, too. Every time a guest came to visit, the first stop on Mother's tour of the house was the sofa where the future Pope sat.

These were magical years in our family, our together years, those charmed days before the world broke out in war. Whether it was in Bronxville on the autumn break from school or in Hyannis Port during the glorious, languid summer, my brothers, sisters, and I occupied the same space. We ranged in age: At one end of the line were the older boys, Joe and Jack, fearless, forceful teenagers soon bound for college. At the other end was the young gang, Bobby, Teddy, and me, eager to keep up. Wedged in the middle was a staircase of girls, Rosemary, Kick, Eunice, and Pat, who linked the two ends of the chain fast together.

At work on Wall Street and then in Washington, DC, Dad was intensely occupied with economic and international concerns. And we younger children could not avoid overhearing his constant dinnertime conversation with the older boys about the conflict simmering overseas. They debated back and forth about what the next step should be for America. Those of us at the small table did not quite grasp the enormity of the threat that occupied their minds and would so soon occupy our entire nation.

Despite the fun, there were still difficult times, and suffering did come close to home. Most profoundly for our mother, she lost her beloved sister Agnes during that period in our lives. The two of them were close in age and had been inseparable growing up. Mother was Agnes's only bridesmaid when she married New York lawyer Joseph Gargan in 1929. When she died just seven years later, Mother was devastated.

Yet from that tragedy came a wonderful addition to our household. Mother and Dad welcomed Agnes's three children, our cousins Mary Jo, Joey, and little Ann, as members of our own family. Their father had a very demanding work schedule, so for the next many summers, our cousins came to stay. The Cape house had plenty of room and activities galore. Mary Jo was close in age to Pat, so they fell right in with each other. Joey matched up with Teddy and became his lifelong pal. Ann was younger than the rest of us, and many years later, when Dad became ill, she became his close companion and caregiver.

Once they arrived for the first summer, the Gargans were a constant presence. Grandpa and Grandma Fitzgerald, who were also their grandparents, frequently visited, and the household was in perpetual motion. Brothers and sisters came and went. Friends joined in from the village or school. We simply added more plates to the table and squeezed in more chairs.

For those few short years under the same roof, before separation and war, our family was together and we were one. These were happy times.

4

Closet Castaways

*I looked on child rearing not only as a work of
love and duty but as a profession that was fully
as interesting and challenging as any honorable
profession in the world and one that demanded
the best that I could bring to it.*

—ROSE FITZGERALD KENNEDY

For each of us, our education began in the nursery. And on days when we were especially naughty, it was reinforced in the closet.

Mother sent us to the closet when she had had enough. In the parental parlance of today, it was where we had our "time out."

"Jean, do not try that again," Mother warned one afternoon,

spying me near the plate of chocolate chip cookies sitting on the kitchen table. But once she was out of sight, I reached for a cookie anyway—and into the closet I went.

Mother had absolutely no patience for children who did not obey. Since there were nine of us—or twelve when the Gargans came to stay—Mother knew that if even one child bucked up against her, the rest of us would surely follow. The last thing she needed was a herd of unruly bucks racing through the house writing on walls, pulling at curtains, and swiping cookies from plates. She made it clear that we were not allowed to have our own way or contradict her. She made decisions "for our own good." It was the only way she could keep any sense of order and peace.

$\mathcal{M}$other was never untucked or untidy. Her hands were always in use, directing, instructing, reading, playing the piano, darning, and jotting down notes for each of us on the index cards in her special file box. When we moved to England in 1938, for Dad to take up his position as U.S. ambassador to the Court of St. James, the British press used her filing system as an example of American ingenuity and efficiency. But Mother always insisted the situation was much simpler than that. The cards helped preserve her sanity and ensure our survival.

Mother also kept up with important events in the news, and she expected us to do the same. Early each morning, after returning from Mass, she sat down in the kitchen with a cup of tea or coffee and the daily newspaper. In Bronxville, she might be reading the *New York Times*; at Hyannis Port she took the *Cape Cod Standard-Times*. If she came across an interesting story or tidbit, she clipped it out, pulled a safety pin from her pocket, and attached it to her dress. In this way she was certain to remember to tell us about it when we arrived at the table for breakfast.

We usually wandered, one by one, into the breakfast room to find Mother already there, still reading the newspaper with these clippings pinned up the side of her dress. As we started in on our toast and cereal, Mother would begin to unpin each article. She would read us the contents and then ask our opinions.

"Children, there are enormous dust storms sweeping through the Great Plains. Those poor people are breathing in dust and soot. What would you do if you were in their situation?"

"Stock up on some masks?" Bobby might venture, in between sips of orange juice.

"Or run into the sea?" Teddy would add.

Then Mother would take up another clip:

"Did you know that Eleanor Roosevelt is bringing bunnies to

the White House? She sounds like you, Bobby! What do you think she'll name them, Pat?"

"Flopsy, Mopsy, and Cottontail," Pat answered assuredly.

"And Peter!" I added.

Then another:

"Poor Amelia Earhart is still missing. It's been weeks. It seems impossible they haven't found her. Where do you think she could be, children?"

"They say she may have landed on the Polynesian islands," Jack surmised.

"I have a stamp from there," Bobby inserted proudly.

"That sounds like it might be fun!" said Teddy.

Our breakfasts were feasts of facts.

Having nine children made Mother supremely practical and efficient. If one of us came down with a contagious illness, it simply made sense to her that the rest of us should come down with it too. Why spend the year cycling child after child through the flu, measles, and chicken pox when we could get it over with in one go? As soon as the doctor stepped from the room of a sibling to report an infectious disease, the rest of us were hustled inside by Mother to play. Within days we all would be scratching or wheezing in unison. It was much more bearable to suffer through it together, and after a week the sickness was out of the house for good.

Of course, Mother did not keep the household moving and all of us educated, fed, and in good health alone. Dad was constantly in the mix, of course, but he was regularly traveling for his work in the financial and film industries, and later as a government official in Washington. So Mother was fortunate enough to recieve assistance from several women who had immigrated to the United States from Ireland to seek work with American families as nurses or nannies. Mother, so conscious of her own great-grandparents' immigration story, was eager to give these women a place in our home. And they were central to her life and to ours.

These resilient women took charge of diaper changing, bottle washing, pram pushing, ear scrubbing, and meal planning. Women did not have the luxury of disposable diapers or electric clothes dryers in those days. Instead, they had to use cloth diapers, which were washed and dried in the open air before being put back on the baby to be soiled again. There was no jarred baby food, either, so our nurses boiled the vegetables and fruit from scratch and pureed them by hand. They helped corral us every morning outside the bathroom to file in and brush our teeth. We repeated the same drill every night. Mother often spoke of her enormous gratitude for the assistance our nurses gave her during those early days, when the tasks of keeping nine children and their needs in order could quickly become overwhelming.

Dad also deeply appreciated the help our nurses gave Mother, and insisted that they use his car if they were going out for the evening. Or he would invite them to watch a movie with the family in the small theater he set up in the basement of the house in Hyannis Port. "Take time for a swim today. The water is wonderful," Dad would tell Keela Cahill, one of our favorite nurses. Or, "Use the symphony tickets this week. I'll be away." He wanted to make sure these women were not treated with the same disdain and disrespect that had accosted his ancestors when they arrived in Boston two generations before.

In addition to Miss Cahill, looming large in our life was Katherine "Kikoo" Conboy, a broad Irish woman with dark hair and a round face that was often jolly and sometimes stern, depending on the day's events. Kikoo was a constant presence in the nursery when Pat, Bobby, Teddy, and I were small. She would pull us in and wrap us in her arms when we were sad or scared. "There, there," she would say. Kikoo instantly appeared with Mercurochrome and a bandage if we scraped a knee or stubbed a toe. But she also had a strong side and had no patience for nonsense. On the days when we stepped out of line, Kikoo was not one to mess with. She took her job very seriously.

*Y*ou bold stump!" Kikoo bellowed out the window at Bobby, who seemed always to be doing what she did not want him to do. "Come inside this instant!" she hollered as he dashed around the fence to play ball in the alley with his friends. Miss Cahill once had to stop Kikoo from banging Bobby's head against the wall to get him to go to sleep. It is still unclear why Kikoo thought banging Bobby's head would do the trick, but thanks to Miss Cahill, we never actually had to find out if it worked.

Every Saturday night, at Mother's instruction, Kikoo would line us up against the wall to record our heights. She would pull out her long measuring stick and place a discreet pencil mark on the door's molding, just in line or slightly above the mark from the previous Saturday. Then she would march us over to the scale to record our weights. If we had gone up or down a few pounds, Kikoo took action. The child losing too much weight would get her clucking sympathy, and perhaps an extra helping of mashed potatoes at supper that evening. But the child gaining too much weight elicited no such sympathy at all. "You bold stump!" she would declare, shaking her head as if we had just committed the lowest of mortal sins. No cookies for the rest of the week.

Which was how I ended up in the closet.

That one extra cookie I swiped off the table was too much for Kikoo, and it was too much for Mother. When Kikoo indignantly reported my transgression, Mother had only one punishment that fit the crime.

"You need to spend some time in the closet, Jean," she said as she took my hand and escorted me up the stairs, through her bedroom, and toward the waiting closet door.

$\mathcal{I}$ walked inside and slunk to the floor. Mother turned and left, pulling the door behind her, leaving only one small crack where the light could peek through. That light allowed me to look around at her clothes, hats, and shoes that surrounded me.

I did not want time out. I wanted time with my sisters and brothers, who were playing ball on the lawn outside. I could hear their voices through the open window in Mother's bedroom. Who was winning the game? Did they just say my name? Do they realize that I'm not there?

Teddy and me in Bronxville (1933)

Then . . . SMASH!

Something had happened. There was a sudden scuffling in the
bedroom. The closet door opened swiftly—a flash of light—a

figure whisked past me, and then all was dark and still again. All except for a small sniffle at the other side of the closet.

I was no longer alone.

I recognized the sound of that sniffle immediately.

"Teddy?" I whispered.

"Who's that?" came the startled voice of my little brother from across the floor.

"It's Jean, silly. What are you doing in here?"

"I broke a window with the ball," he replied. "Mother called me out of the game and put me in here. What are *you* doing here?"

It was so unlike Mother to have made a mistake like this. But she had. Mother, who never seemed to forget anything, had forgotten that I was already in the closet, and she had put Teddy in here, too. Teddy, my playmate. Teddy, my loyal chum. How perfect! My luck had changed.

We sat in silence for a minute or two as Teddy continued to sniffle and rub tears from his cheeks.

"Teddy?" I finally whispered again. Through the faint light, I saw his head turn in my direction.

Boldly, I stuck my grubby toes into Mother's new ivory satin shoes. If she had been there to see me handling them that way, she might have collapsed. But we were blissfully on our own.

"Do you think these heels go well with my dungarees?" I asked.

I saw something shine across the closet: Teddy's eyes. His tears had stopped.

I started popping Mother's hats on and off the top of my head.

"Do you prefer the red chapeau with the blue feather or the yellow chapeau with the green ribbon?" I asked.

Teddy was moving closer. A smile started taking over his face.

I rolled onto my back to stare at the ceiling. He rolled onto his back too, his head next to mine.

"Teddy," I said, after a minute or two, "do you think we will ever get out of this closet? Do you think we'll ever be set free? Or do you think we'll be like Amelia Earhart, stuck here forever?"

He giggled. It had worked. Teddy was himself again—and that meant he was ready to play.

"Where do you think she is, Jean? Amelia Earhart?" he asked, then continued, without taking a breath: "She must be on a deserted island, don't you think? Doesn't she have to be alive? I can't believe they can't find her. Don't you think that's odd? Maybe she'll come home, but I'm not so sure. It's such a mystery, Jeannie. I just love a good mystery!"

So went the rest of our time in the closet. Teddy and I sailed off to a faraway island together, to the bright blue world where

Amelia Earhart must have landed, waking on the sand and blinking in the sun, ready to make a new home for herself.

We imagined how afraid she must be so far from home, but also how intrigued at the curiosities around her. We shivered at the strange noises of the jungle. We walked miles down the sandy beach with her looking for traces of life. We made friends with chimpanzees and ate plump berries and coconut.

Lying on our backs, looking around at Mother's sparkling gowns, we imagined the moon gleaming down on us with a cockeyed smile. On our island, the waves splashed, and the birds cawed. Amelia was content, and so were we.

All too soon, Mother returned.

"Teddy, dear," she said as she opened the door. Then her eyes fell on me.

"Oh Jean. You're in here, too"—and she realized in that instant what she had done. A brief spark of recognition passed over her face, but nothing more. Mother was not about to admit her mistake. She knew that if the story got around to the others, she would never live it down.

She looked quickly across the closet, making sure her dresses, hats, and shoes were intact. Satisfied that we had not torn the place apart, she pardoned the criminals.

"Your time is up, you two," she pronounced. "You can go out and play again."

Teddy and I rose to our feet and stretched. Then, stepping out of the South Pacific, we walked past Mother and returned to Cape Cod.

5

Faith, Values, and
Hard Work

*One can't walk out on a job just because one
wants to. Obligations must be fulfilled.*

—JOSEPH P. KENNEDY

*M*other and Dad made sure we understood that we
were not the center of the universe. Rather, we were here to serve
a greater good. Our faith, and our relationship with God, was extremely important to our parents. For Mother in particular, faith
was the bedrock of her life. She and Dad believed that a strong
ethical and spiritual compass was essential to our leading meaningful lives. And as their parents and grandparents had before
them, they turned to the Catholic Church and its teachings as our
guide for what was right and wrong.

Faith was not an oppressive presence in our house, but it was a constant one, a buttress and a buffer against life's troubles. From a practical point of view, Mother made certain that we observed all the important rituals, like most Irish-Catholic families in the early days of the twentieth century. We were baptized within days of being born to ensure that we were cleansed of original sin. The sacraments of First Confession, First Holy Communion, and Confirmation followed as we grew.

As little children, before and after meals and right before bed, Mother would take our small hands in hers and guide us in the words that served as the introduction to prayer for the rest of our lives. Placing our fingers first on our foreheads, then on our chests, then to the left shoulder, and then to our right, she had us repeat after her, "In the name of the Father, and of the Son, and of the Holy Ghost."

Mother taught us all the prayers necessary for a good Catholic upbringing: Hail Mary, Our Father, Hail Holy Queen. We learned to recite the rosary, and she always insisted we had one in our pockets when we traveled. Without fail, we attended Mass every Sunday, on First Fridays, and on the Holy Days of Obligation—the Epiphany on January 6; the Assumption of the Blessed Virgin Mary on August 15, the Immaculate Conception on December 8, and so on. It was against all family and church rules even to think of missing Mass on those designated days. My brothers served as altar boys at our local church. We girls wore veils on our heads and carried prayer books, genuflecting in the aisle and then filing into the pews side by side for the solemn Latin service. All other Catholic families we knew were doing the same. It was not unusual.

It would have been easy, in the midst of all this ritual, for Mother and Dad to forget the point of it all. But they made certain that our true religious education remained focused where they felt it needed to be: on faith and service rather than on structure and dogma. A favorite evening lesson of Mother's was the Beatitudes, taken from the Gospel of St. Matthew. Mother would recite them for us, pausing, for emphasis, at the end of each phrase. It was one clear message, delivered regularly to nine sets of attentive ears.

> Blessed are the poor in spirit, for theirs is the kingdom of heaven.
> Blessed are they who mourn, for they shall be comforted.
> Blessed are the meek, for they shall inherit the earth.
> Blessed are they who hunger and thirst for righteousness, for they shall be satisfied.
> Blessed are the merciful, for they shall obtain mercy.
> Blessed are the pure of heart, for they shall see God.

Blessed are the peacemakers, for they shall be called children
   of God.
Blessed are they who are persecuted for the sake of righteous-
   ness, for theirs is the kingdom of heaven.

The lessons of the Beatitudes became the primer for how
Mother and Dad felt we should live. They became the guide for
how they felt we should treat one another and treat other people.
Often after reciting them, Mother would remind us, "These are
the people whom He wants you to help."

She and Dad believed we could serve the world most fully by
engaging in what Dad called "good works, good reputation, and
hard work." "Honest effort sometimes is disappointing in its re-
sults, but in the long run never misses," Dad wrote to a teenage
Jack. Dad and Mother did not abide idleness nor did they listen
to complaining. And we were certainly expected to tread a prin-
cipled path.

"Without meaning any criticism of your very excellent character," Dad once wrote to Kick, "I have noted that with you popular
opinions are frequently accepted as true opinions. There is nothing particularly wrong in this, it's safe and you've got plenty of
company, which you like. But I think you'll find that the majority
are only occasionally well-informed and that your own judgement

is frequently better, and will always be more Christian than opinion in mass. So don't bum rides on other people's opinions. It's lazy
at best and in some cases much worse."

It was a pointed warning and typical of the kind Dad delivered when he feared we might veer off course. No doubt he had
been observing Kick, his fun-loving daughter, swirling in and out
of the house with friends. Each weekend, she piled into a car with
girls from school, and their laughter would trail them out the
drive. Dad and Kick were immensely close. He admired her knack
for instantly drawing people in. Yet he also wanted to remind her
to think and act for herself.

I suspect it was because their parents came from so little themselves that Mother and Dad placed such great store in good character and hard work. Mother's father, John Fitzgerald, had to
raise and support his brothers when his Irish immigrant parents
died quite suddenly. Luckily his love of the common man and his
natural gregariousness led him in the direction of politics, where
he achieved enormous success. When he was not holding office,
Grandpa was hard at work at the helm of a weekly newspaper he
purchased called the *Republic*.

Mother recalled how Grandpa often left the house in the morning before anyone else was up and would return late, after dinner.
Where other men might have collapsed into a chair, he would pace

the room, regaling his wife and young children with stories of the people he met on the train or his current fight for the little man in City Hall. To Grandpa, service to others was a joy.

Dad's father had similar roots, and the same outlook. I did not know my grandfather P.J., because he died before I was born. Still, I often heard Dad speak with tremendous pride about his father and what he had achieved. When P.J. was a very young child in East Boston, he saw his own father die, nearly penniless. His mother, Bridget, opened a shop where P.J. worked running errands and making deliveries along the docks, to support her and his three siblings. Because of their desperate circumstances, P.J. never finished elementary school. He saved enough not only to support his family, but to help his neighbors with loans. He also became involved in politics and pursued and held elected office, though, unlike Grandpa Fitzgerald, he preferred to stay behind the scenes.

P.J. was Dad's role model for how to persevere in business while also being a good man. When Dad entered the workforce himself, he followed in his father's footsteps. A young man with a steely, determined air about him, Dad struck out into life to succeed. Even at eight years old, he was a go-getter. He loved to read and was very excited when the Larkin Soap Company announced it would award a beautiful oak bookcase to the person who gathered the most tickets enclosed in its products. His sister, our aunt

Loretta Connelly, told us that Dad sold soap to everyone he knew and then asked if he could have the ticket inside. He organized a team of friends to span out across East Boston and do the same. It was no surprise when Larkin awarded the bookcase to a young Joe Kennedy.

Even as a child, Dad was never not working. He sold newspapers to the workers on the docks in East Boston and candies and peanuts to passengers on the boats. He was one of the few Irish-Catholic boys to be admitted to Harvard College at that time in history, and he excelled there. His industrious spirit ever on display, Dad partnered with a school friend to buy a bus that they used to shuttle students to and from school, earning quite a healthy sum to support their education. Clearly drawn to economics and finance, after he graduated in 1912, Dad took his first job as a state bank examiner. A year later, in 1913, during the time he was seriously courting Mother, he was called upon by his father to help save the Columbia Trust Company, a bank that P.J. had helped found and that was the target of a takeover. After a very public battle, Columbia Trust remained independent and Dad was chosen as its leader, becoming the youngest bank president in the nation at the age of twenty-five.

Dad continued his steep rise in business, making shrewd and, what some considered at the time, risky investments in finance,

Dad at work

publishing, and film. He had a well-earned reputation for being a tough, skillful negotiator. It was a reputation I suspect he enjoyed. Yet he was enormously bighearted and openhanded in a way that few people would ever imagine and others would never acknowledge. Ever loyal and devoted to the Christian Brothers who educated him beginning at the age of six, Dad continued to quietly support them for decades. Scores of friends and strangers reported to us similar acts of kindness over the years. Like his father before him, Dad worked behind the scenes.

"What interests me is that he has done so many generous things for so many people without wanting such acts to be known," Dad's friend Carroll Rosenbloom once recalled.

I suspect Dad's generosity grew from his desire to always remain self-aware and self-effacing. He was an example to us that no matter what good fortune came our way, it should not be accompanied by pride. On his desk, Dad kept a poem, "The Indispensable Man," by Saxon White Kessinger, which served as his reminder to himself of his place in the world:

Sometime when you're feeling important
Sometime when your ego's in bloom
Sometime when you take it for granted
You're the best qualified in the room

Sometime when you feel that your going
Would leave an unfillable hole
Just follow these simple instructions
And see how they humble your soul;

Take a bucket and fill it with water
Put your hand in it, up to the wrist
Pull it out . . . and the hole that's remaining
Is a measure of how you'll be missed.

Dad's business was constant and exhausting. It often called him far away from home. We missed him terribly every time he was away, and sent him letters and telegrams telling him so: "Dear Daddy: I hope you come home soon. I think of you every night. I want to kiss you," wrote a five-year-old Pat. He missed us right back, and wrote constant letters with news from his travels and questions about our lives at home.

"I hope the skating turns out to be lots of fun and I will be anxious to hear just what happened to you," he wrote Rosemary. "Be sure to wear a big pillow where you sit down so that when you sit on the ice (as I know you will) you won't get too black and blue."

Every time Dad arrived back home, he gave each of us a huge embrace and a kiss on the cheek. He flooded us with affection, even as we grew and went away to college. Less confident young men might have been embarrassed to see their Dad striding toward them across campus, smile wide and arms extended. But Joe and Jack just hugged him right back.

"He wasn't around as much as some fathers, but . . . he made his children feel that they were the most important things in the world to him," Jack later said. Dad was not a bystander in our lives. He was our champion and defender. He was there urging us on from the sidelines of our games and from the shores of our races.

"After you have done the best you can," he told us, "then the hell with it." But if we had not pushed ourselves to the limit, he told us so, and firmly. It was not easy to take but always essential to hear.

Mother often reminded us how much Dad loved us and that he showed his love by working so hard to ensure our security and future. She, in the meantime, ensured that the household ran efficiently and that every penny Dad earned was spent wisely. She never wanted us to take money for granted, and she didn't, either.

"I grew up with the idea that one should be careful with money, that none should be spent without good and sufficient reason—tangible and intangible—to justify each expense," Mother later wrote.

Like her mother before her, Mother darned socks and mended trousers when her family was very young. And for her entire life, she replaced a chair cushion or discarded a dress only if it was badly worn.

She could drive you crazy on the golf course looking for balls that had been hit in the rough. Balls were too expensive to lose. If you had three balls when you began the game, you had better bring three home.

We received presents just two times a year, at Christmas and on our birthdays, and they were never abundant nor extravagant.

One Christmas might bring a wind-up toy or a new sweater. We did not get a new bicycle each year like some might imagine. Bicycles were meant to be ridden until they fell to pieces. If a sled had a bad encounter with a tree, we hammered it back together and bound it with twine. Nothing went to waste. Wastefulness was a sin.

Teddy often told the story about when he went away to boarding school and wanted to take his bicycle along with him.

"How many other boys have bicycles at school?" Dad asked him when Teddy announced his plans.

"Only a handful," Teddy replied.

"I think you can get along without yours for a few months like everyone else," Dad concluded.

Dad could not bear the idea that his children might ever come across as spoiled or accustomed to special privileges.

"In looking over the monthly statement from Choate, I notice there is a charge of $10.80 for suit pressing for the month of March," he wrote Jack at school. "It strikes me that this is very high and while I want you to keep looking well, I think that if you spent a little more time picking up your clothes instead of leaving them on the floor, it wouldn't be necessary to have them pressed so often."

Mother and Dad insisted that we take care of our rooms, making our beds each morning—"No sloppy corners!"—and picking up our toys and clothes from the floor. As we grew older, our chores extended throughout the house. I spent many afternoons attempting to achieve the unachievable goal of sweeping all the sand off the front porch, while one of the boys was tackling the lawn with the mower. Mother and Dad also sent us out to work in the world. "If you want something, you have to go out and get it," Dad would say.

Teddy mucked stables at the barn in Osterville, and at Mother's suggestion, I began to volunteer during the summers at the hospital in Hyannis. I made the beds, changed bedpans, rolled bandages, swept the floors, and helped the nurses with their various responsibilities. It made me want to be a doctor, and I told Dad about my plans. "No, women aren't doctors," he responded. "That's not a suitable profession for a woman." Those were the days in which we lived, when women were viewed largely as homemakers and were not encouraged to pursue the professions traditionally held by men.

Dad and Mother wanted us to understand our obligations. We were lucky to have the things we had. We were lucky to be healthy, with sharp brains. Not only should we not whine; we also had to give back. We had to use our talents and our minds. We must give our all.

One day, while playing football on the lawn, I got my feelings hurt. My brothers would not pass me the ball. I left the game in a huff, stomped up the front stairs and into the hall, where Mother was gathering her golf clubs and heading out to play.

"Whatever is the matter, Jean?" she asked.

"They aren't being fair," I huffed.

Bristling at my whine, she delivered a swift instruction: "Go up to see your father."

This was not what I wanted to hear. I would rather have gone to my room to sulk alone. But Mother had spoken, so I slunk up the stairs and positioned myself outside my father's door.

"Dad," I said, to capture his attention.

Busy at his desk, he peered over the papers he was reading and caught my eye.

"Why aren't you outside in the game, Jean?"

We could hear the others, below on the lawn, just outside his window.

"They're not fair," I burst out. "They won't pass me the ball. They won't even look at me. I hate football, and I'm not playing. I quit!"

Jack, me, Mother, Dad, Pat, Bobby, and Eunice, with Teddy in front
(At the Cape, 1948)

Silence. Dad took a minute to let the words fill the room. *Not Fair. Hate. Quit.* They bounced off the ceiling and landed back down in my ears.

"Jean," Dad said, "do you see what you've done?"

I looked at him. I had no idea what he was talking about.

"Your team needs you. They're counting on you. If you're not there, they have one fewer person on their side.

"The chances are greater that they will lose," he continued. "There will be a hole, and someone will throw the ball and make a touchdown. It is very tough to have an extra player on the other side."

In all my self-pity and anger, it had never occurred to me that I might be letting the others down. Sure, the boys might be the stars of the game, but I was needed to make it even.

"Head back out there, Jean," Dad said. And there was no question that I would.

"But first," he said, grinning, "let's have some butter crunch."

Had I heard him right?

Dad always claimed he was on a diet. He would make a tremendous scene of turning down dessert every night after supper. A sad look would come over his face as if he were turning down the Crown Jewels.

"Oh no. I'm not touching it!" he would pronounce, raising his hands high to ward off the ice cream and profiteroles that beckoned from the table. Then he would look around the table at each of us, beaming with pride at his self-discipline.

It was quite a show. But we all knew the truth. Upstairs, in his closet, tucked in between his cashmere sweaters, Dad kept a secret stash of Katie Lynch's Butter Crunch for clandestine late-night snacks.

None of us dared mention it, but we were certain it was there. And now here was Dad, offering it to me. Here was Dad, walking out of the closet and holding out the box of precious candy in my direction.

I pretended to be surprised.

"Where did you get the butter crunch, Daddy?"

Then I hopped up on the edge of his bed and he settled down next to me.

There we sat, together, my generous, princely father and me, nibbling the forbidden treat. The world was fair again.

6

Grandma and Grandpa Fitzgerald

I think there is truth in the idea that "opposites attract," because it was certainly true in the case of my parents. Father was an extrovert; Mother was innately rather shy and reserved. He would talk with anybody about anything. When she spoke, it was usually directly and to the point.

—ROSE FITZGERALD KENNEDY

My grandma rode to school on a horse. Even as a small child, I found that fact hard to grasp. She would mention it in passing, as if it were just a normal day of any child's life growing up. To Grandma, it was not an incredible event or an adventure. But to me, it was exotic, fascinating, and glamorous. To think that just a few decades earlier, my grandma, Mary Josephine Hannon, had been a little girl just about my age and size, living in Acton, Massachusetts, a small town north of Boston. She rose for school, just like I did. I am sure she ate breakfast and combed her hair just like I did. But then our morning routines took very different paths.

While I jumped into a car or onto a bicycle for the quick trip to school, I thought of Grandma and how differently she made the same trip. In my imagination, she wore a simple poplin dress that was neatly pressed. Her hair was pinned up atop her head. Grandma would head to the barn, where she nuzzled her mare, gave her a handful of oats, and carefully groomed her chestnut coat and mane. Once saddled up, Grandma would mount the horse. Then, positioned ever so gracefully on the sidesaddle, her two legs swung to one side, she would trot off to school. How was it possible that the world had changed so much since my grandma was a child? But then, her life was so different from the lives of her parents in Ireland, who most likely had no breakfast or horse at all and instead struggled to find food.

Mary Josephine Hannon, my mother's mother, was a beauty. Petite, with a trim figure she was very careful to maintain,

Grandma and Grandpa, with Mother at the piano, at the party to celebrate their fiftieth wedding anniversary (Boston, 1939)

Grandma had long, wavy dark hair that she wore wrapped up in a bun, as was the fashion of her day. She was elegant and other-worldly. I thought she belonged in another age, in a grand palace in some far-off, exotic land for dark-haired, radiant beings.

She and my grandpa John Fitzgerald were devoted to each other. As we often heard, the happiest day of Grandpa's life was the day he married Grandma. "The first time I met her, I knew," he said, beaming his signature Grandpa grin. Every time he looked at her, I could tell that Grandpa was still seeing the young, beautiful girl he first clapped eyes on decades before. Dad was enchanted by his mother-in-law as well, and appreciated her calm and grace.

Grandma was a woman of habit, a woman of routine, who had no need or inclination to be out in front. That was Grandpa's job. Grandma was steadily in the background, managing the household and ensuring that her children were raised well, with ever-broadening minds and deepening faith. She was not boring by any means nor was she stern. The word that comes to mind for Grandma is lovely. Grandma was gentle, she was graceful, she was quiet.

Grandma found her opposite in Grandpa. He adored the lime-light that she avoided. He was made for it. He came into this world ready for action.

Grandpa was born in 1863 in the North End of Boston, in a

neighborhood inhabited by many other Irish immigrant families of that time. His family lived in the shadow of the Old North Church, where less than a century before two lanterns blazing in a belfry had called a fledgling republic to revolution. Grandpa was born into another turbulent time in America, at the height of the Civil War, which had drawn tens of thousands of his fellow Irishmen to fight for the Union a thousand miles away to the south.

The first air Grandpa inhaled was revolutionary. He learned to walk on the cobblestone streets of history. He explored the same alleyways as Paul Revere and his compatriots. Perhaps he adopted his fervor for life from their spirit. Or perhaps his zeal harkened further back in time, to the dirt roads of Ireland, where his ancestors also had to fight for their freedom.

Whatever the reason, Grandpa simply had a passion for life. He was a joyful, buoyant fellow whom everyone loved. Grandpa had an energy that was at the same time contagious and exhausting. By the time I was born, he was an established legend, both in Boston history and in the history of our family. He was the beloved and respected "Honey Fitz," former U.S. congressman and Boston's first Irish-Catholic mayor. "The greatest mayor Boston ever had!" Joe and Jack would pronounce to their college friends whenever Grandpa's name was mentioned.

An Irish showman by nature, Grandpa was known for bursting

into song at a moment's notice. He often opted to sing a tune rather than deliver a boring speech, a trait the people of Boston greatly appreciated. In fact, they nicknamed him Honey Fitz because of his melodious singing voice. And he found the perfect accompanist in our mother.

Grandma was not fond of politics, so Mother took her place at Grandpa's side on the campaign trail as a teenager. Grandpa's signature song was "Sweet Adeline," a modern tune at the turn of the century that became ever linked to his name. He would sing it at the drop of a hat—often screaming the words so the crowd could hear. Highly accomplished at the piano keyboard, Mother accompanied him lightly as he sang. She kept a natural pace, slowing down or speeding up to match his tempo, pausing momentarily for his impromptu comment or quip. "You're the flower of my heart, Sweet Adeline . . . My Adeline . . ."

Acting upon his campaign slogan "A bigger, busier, and better Boston," Grandpa transformed the city and drastically improved its commerce during his two terms in office. Some of Boston's institutions were born under his watch. The Franklin Park Zoo, Filene's department store, and the Copley Plaza Hotel opened their doors. Grandpa convinced the Massachusetts state legislature to invest nine million dollars in revitalizing the city's struggling port.

Yet to the nine of us growing up, his most important contribution to Boston was the towering Christmas tree that lit up Boston Common every December. Always a man of the people, Grandpa was the first mayor in the United States to erect a Christmas tree in a public park so that everyone could enjoy it. We thought it was marvelous, particularly after we learned that he beat New York City to the punch. He would tell the story of Boston's first Christmas tree at random times throughout the year, but without fail on Christmas Eve, when he painted a picture of the citizens of Boston, citizens of every shape, size, and color, making their way to Boston Common for the event.

They *ooh*'d and *ah*'d at the enormous tree that, in an instant, was set aglow. Then Grandpa, taking a deep breath into his broad chest, burst out in song for the crowd. "Oh Christmas Tree, Oh Christmas Tree . . ." Most certainly the snow was falling on that storied evening, and thermoses of hot chocolate were passed among the huddled crowd. At least that is how I imagined it while I listened to Grandpa recount the long-ago evening. As he wound down his story for us, that was Mother's cue to rise from her seat on our sofa and take her place at the piano in our living room. Grandpa then rose to stand beside her and began to lead all of us in Christmas carols just like he had on that crisp winter night for the people of Boston.

"Why did you do it, Grandpa? Why did you put up the tree?" we would ask him.

"Everyone deserves a Christmas tree," he replied, simply.

We also relished hearing Grandpa tell of the day in April 1912 when, at the age of forty-nine, he threw out the first pitch for the inaugural game in Fenway Park between his hallowed Boston Red Sox and the New York Highlanders. My brothers were especially keen for him to tell the story again and again and again. For the week before the game, Grandpa would explain, Bostonians had been distressed and distracted with the news that the mighty ocean liner *Titanic* had gone down in the North Atlantic, its hold jammed with poor Irish immigrants making their way to America. So when Opening Day arrived, the people were ready for a reason to relax. Grandpa and his friends in the "Royal Rooters" fan club were among them. They had been waiting for this day since the first brick was laid a year before.

Opening Day was originally slated for April 17, but agonizingly for Grandpa, it rained for three days and the game was postponed. So when the sun finally rose brightly on Saturday, April 20, Grandpa was primed for the game. In top hat, tie, and a three-piece suit, he stepped to the mound, and a hush of anticipation fell on the twenty-four thousand who had gathered there.

He clenched the ball, wound up, and, as he always told it, "threw the perfect pitch!" The crowd roared, the game began, and away Grandpa's story went, play by play. The Red Sox started out slowly; they could not find their pace, and the Highlanders advanced. But as the innings progressed, the home team got its footing on its new turf and began to rally. In the final inning, the Red Sox squeezed by the Highlanders 7–6. "Victory!" Grandpa bellowed. My brothers, who followed Grandpa's words so closely they might as well have been in the stands, cheered, "Yes!" Then they pleaded, "Grandpa, tell it again." They did not have to ask him twice.

Despite Grandpa's success in life, and the ease with which it seemed to come to him, Mother always reminded us how hard he had worked to achieve what he did. Grandpa was small in stature (just five foot, seven inches tall), and that seemed to only fuel his gumption rather than deter it. He devoured learning from his earliest years. No topic was beyond his interest: poetry, science, language, engineering. It always said something to me about his intellect and versatility that this future mayor first wanted to be a doctor. He was accepted to Harvard Medical School, and I am sure he would have gone on to be a great physician had fate not played a hand when both his parents, still young adults, died within a year of each other. Grandpa dropped out of medical school to take care of his younger brothers.

Maybe it was because he could not finish college that he read everything he could get his hands on, starting with his beloved Boston papers, from which he cut out interesting facts and stories with his penknife so he could pass them on to others to absorb. Yet of all the topics he enjoyed, history—most particularly the history of Boston—was his first love. No one knew it better or could recall it with such a dramatic flair. When any of us visited Boston, he would take us walking along the streets, supplementing our often stiff classroom lectures with his colorful accounts of the living and breathing history all around us. My brothers who attended Harvard were particularly lucky because Grandpa was just down the road and regularly arranged weekend engagements.

Grandpa would meet them, perhaps in the lobby of the Bellevue Hotel, and rush to speak: "When in the course of human events it becomes necessary for one people . . ." he would bellow.

And Joe or Jack would immediately jump in to take up the challenge, shouting the words over Grandpa while bursting with laughter:

" . . . to dissolve the political bands which have connected them with another and to assume among the powers of the earth, the separate and equal station to which the laws of nature and of nature's God entitle them, a decent respect to the opinions of mankind requires that they should declare the causes which impel them to separation!'"

In that way, Grandpa taught them to memorize the Declaration of Independence.

While we were growing up, Grandpa was often in the center of things. Though his visits were not long, he took up more than his share of space in the room. He was there at Thanksgiving in Bronxville, declaring that the turkey was the biggest bird he had ever seen. He was there in Hyannis Port, telling stories on the porch, newspapers scattered around his feet.

And he was there in London, settling in for a visit soon after our family moved to the imposing Embassy Residence at 14 Prince's Gate, just off Hyde Park. Each afternoon he would retire to his bedroom for a rest. But after a few days, word came down from the fourth floor that he was not resting at all. Rather, Grandpa was composing letter after letter to his friends back in Boston, urging them to set sail for London, too.

"Come visit! Rose and Joe have plenty of room," he wrote on the official embassy stationery, beneath the engraved gold seal of the United States.

We were never sure if he actually mailed them, but no one ever showed up!

Grandpa was a character of the first order. He could not help but laugh at his own jokes. In fact, he often had trouble getting through a joke because he was laughing so hard that he had to stop telling it. We children would get frustrated because we would forever hear the set up but we rarely heard the punch line. It seems the only person who never seemed the slightest bit bothered by Grandpa's endless gaiety was Grandma. She understood him so well, and always enjoyed his show.

Later on in our grandparents' life, Mother and Dad purchased a small cottage for them up the road in Hyannis Port, just before the turn into town. True to her reserved nature, Grandma spent most summer days in the quiet of the cottage, alone with her thoughts and prayers.

On summer afternoons, at around 4:00 or 5:00 p.m., in those slow hours in the day between swimming and supper, I would ride my bike down to visit Grandma at the cottage. She greeted me with a soft kiss at the door, dressed in a simple cotton shift, her hair swept up from her face as always. Tea and sugar cookies might be waiting on a china plate on the table, in anticipation of my arrival.

Together we would move to her small sitting room, where we would settle into comfortable chairs and I would read to her, from the newspaper or from a novel we had chosen. I read the gospel

from that morning's Mass or a story from the *Lives of the Saints*. I
read from the plays of Shakespeare or a poem by Yeats:

> *Had I the heavens' embroidered cloths,*
> *Enwrought with golden and silver light . . .*

And Grandma would say, "That's lovely dear, simply lovely."

The afternoon passed. And my quiet grandmother, who
started her life riding to school on a horse, who sustained and
supported the most powerful man in Boston, who lived to see her
grandchildren chart their own course in history, who was the
peaceful presence in our family swirl, would remove her glasses,
close her eyes, nod her head softly, and listen.

7

The Ocean in Our Veins

Teddy, if you leave with the boat,
you come back with the boat.
—JOSEPH P. KENNEDY

*I*n our family, we found common ground on the sea. It did not matter what age you were, and it did not matter if you were a boy or a girl. One of the great advantages of sailing is that everyone on the crew is responsible, everyone has a role to play.

The sea was where we spent most summer days. Even the youngest of us was put to work. As we heaved the ropes, as we tacked into the wind, we learned how to compete and how to cooperate.

All my brothers were serious sailors, and they did not suffer fools on the water. They loved to compete in boat races, and they loved to win. It seemed as if there were endless regattas to join on those bright summer days. The boys raced in the senior division, most often in Jack's twenty-five-foot senior sailboat, *Victura*, which Mother and Dad gave him when he was fifteen.

Eunice raced as well, and matched the boys skill for skill. While Pat and I were more casual sailors, Eunice had a steely focus and determination. Our brothers greatly admired her skills and welcomed her as part of the crew.

An unsuspecting sailor on another boat might have looked over to *Victura* at the start of a race, spied Eunice's young, wiry frame at the tiller, and thought they had a chance. But they would have been sadly mistaken. At the starting gun's report, Eunice would spring into action. In full command, she issued orders left and right as those on the crew scurried to keep up.

"Now! Ready about!" The boat would surge forward, propelled by her will. Eunice did not like second place.

"We received alot [*sic*] of prizes for sailing . . . mostly due to Eunice," Bobby reported in a letter to Dad after one particularly successful summer on the sea.

I often hoped that one lucky day, I would join the boys and Eunice at the helm, maybe even as captain. From that position of power, I would have full permission to order my brothers and sisters around. But that day did not come. Sailing was too important a business in our family to put it in the hands of a daydreamer.

Bobby and me (Hyannis Port, 1931)

Instead, I was given the job of ballast. It was not a job for the faint of heart. In retrospect it was great fun, but at the time it could be terrifying, especially when the mast swinging toward your head was intent on decapitation. Yet there I was, in sunshine or storms, crouched down in position on the bow of the boat.

As the ballast, my role was to balance out the boat with the weight of my body so she would stay steady and not tip too far. I would lurch back and forth across the deck, sucking in saltwater, clothes drenched, ducking the heavy mast that heaved overhead. If I raised my head even an inch, the mast would have taken it right off.

"Jean, move!" my brother would yell, demanding that I scurry faster and faster.

"Starboard!" came the command, and I would fling my body to the right. "Port!" and I would fling to the left.

"Coming about!" I would fall to the deck just as the mast bore down on me.

A quiet day at sea it was not. Yet I wanted to be with my siblings regardless of the circumstances. I suspect that sometimes they wanted to throw me overboard. And they probably would have, if it did not mean having to turn around and pick me back up. The rules of the race required that you go out to sea and return with the same people and items on your boat. Fortunately that rule saved me a number of times—but it did not save Teddy.

One day, Joe became so put out with Teddy for not pulling out the jib fast enough that he lifted him by the seat of his trousers and threw him into the water. Of course, Joe immediately dove in and pulled our little brother back onto the vessel. He would never seriously risk Teddy's life—nor the race!

As we grew, Teddy and I began to compete in races of our own in the junior division, and it gave us no end of pride to report back to our brothers when we finally brought home the top prize: a sterling silver box. "Joe tells me you're leading the Junior One-Design Sailing Class at Wianno," one of his Harvard friends mentioned to me in passing during a visit home. It was a big moment of pride.

It was a tradition for our family that the moment we arrived at Hyannis Port each summer, before we even entered the house, we would run down to the breakwall to say hello to the sea. The breakwall stretched out into the water in front of the house. It buffered our yard during high waves and high winds, including the deadly hurricane of 1938, which we rode out in the basement while the older boys kept running up and down the stairs making sure the house was still in one piece.

This constant battering had taken its toll on the wall over the years, and it was shaky and precarious in spots, rocks nudged loose by the water and winds. That made scrambling up on top of it all the more exciting. Our arms extended for balance, we carefully tested each rock with the toes of our sneakers, risking a tumble with every uneven step. Finally reaching a stable place, we stopped and together waved out to the water. "We're back!" It was our way of guaranteeing good luck for the summer.

It was all around us and within us, the sea. The Nantucket Sound was practically in our backyard, just beyond the lawn. It

Teddy and Jack (Hyannis Port, 1946)

was our playground. Sand lingered in our shoes, on the porch, and in our hair, even after a good washing. We kept close watch on the winds and the tides and the currents. On the Fourth of July, after a long day of racing, we would gather on the sand at the West Beach Club to see the water light up from the fireworks above. When night finally fell, with our windows open in the dark, the sea lulled us to sleep.

Saltwater was in our blood, in our genes.

Our Irish ancestors had great respect for the waters that encircled their island home. The sea could be the Irishman's fickle friend. Legend tells us that in the first century AD, the sea was the only force strong enough to keep the Roman Army from invading Ireland's shores. Yet some six hundred years later, the same waters ushered in the Vikings, who pillaged the island of Ireland and changed its landscape forever.

"Time to put off the world and go somewhere and find my health again in the sea air," wrote Yeats.

Mother certainly inherited the Irish affinity for the sea, and never missed her daily dip during the summers in Hyannis Port. She went out each day before lunch, and we often joined in, swimming alongside her, imitating her graceful strokes. We then would gather on the porch for a lunch of tuna fish sandwiches and milk, looking out at the waves.

Other days, we would pack a picnic and sail as a family out to the islands. If we were game for a long outing, we headed toward Martha's Vineyard or Nantucket, which, even in those days, were teeming with tourists and taffy shops. Shorter stints took us to Great Island, which was closer to our house and less populated. The sailboats did not have motors, so we went out only when the winds were right. There was nothing worse than baking in the hot sun on the deck of a boat in still, listless water, waiting for the sail to catch an elusive breeze.

We sometimes sailed together on one boat, but our convoy would grow to two or three boats if we had friends along. Each of us would take turns, hanging on to a life preserver that we attached to the back of the boat with a long rope. The sailboat would drag us along through the water. Sometimes, when we got close to Great Island, we would let go of the life preserver and swim into the beach, where we played ball or hunted for interesting rocks and shells. Those on the sailboat would drop anchor and then bring a float onto shore with a basket of food. We would spread out our picnic blankets on the island, unwrapping our boiled eggs and sandwiches and popping open our sodas. Then, after a rest, we would head back to the boat.

Even when the sun went down in Hyannis Port, the sea beckoned. "Let's take a walk," Mother would say after the supper dishes had been cleared. We would walk with her up the drive and down the road where the boats from the town were moored on both sides, then up onto the pier. The *whisk, whisk, whisk* of the water hitting the pilings was soothing to hear. Gulls drifted easily overhead.

Mother would lead us to a small shelter at the end of the pier where she would stop to sit on the bench inside. There, we ended our day, side by side, talking quietly in the lamplight, tasting and devouring the familiar salt air.

The last one out of the water each night and the first one in each morning was Teddy. Like all the boys, he was viscerally drawn to the sea, and it was a passion that lasted his entire life.

When our family moved into the house in Hyannis Port, Teddy was not born yet, so there were only ten of us: Mother, Dad, and eight children. Consequently, when Dad purchased our first sailboat, he christened her *Tenovus*.

A few years later, Teddy came along, and Dad purchased another boat. That boat's name? *One More.*

It was probably at that moment that the die was cast for Teddy. As a boy, he had no need for a pet dog or cat. *One More* was his faithful companion.

Teddy learned early in life that a boat is a great responsibility. As he recounted in his memoir, *True Compass*, one rainy day he set out for a sailing excursion with our cousin, and his best pal, Joey Gargan in *One More*. They spent a miserable night on board and come morning, they anchored the boat, swam ashore, and called Dad's chauffeur for a ride back to the house. Walking in the door, drenched and tired, Teddy passed right by Dad and headed up the stairs. I was standing at the top looking down.

"Teddy, I thought you were going for your cruise," Dad said.

"But it was cold! It rained," Teddy replied.

He was halfway up the stairs when Dad's voice stopped him short.

"Teddy, where is the boat?"

There was a pause. I waited to hear what Teddy would say.

"It's anchored. We'll go back and get it later."

"Teddy," Dad said firmly. "If you leave with the boat, you come back with the boat."

The words hit their intended target. Teddy understood what Dad was saying, and he could not believe it. I could not either. But Teddy knew better than to argue. Dad had made his statement, and there was no challenging him, especially when Teddy knew he was right. Soggy and sagging, Teddy did an about-face on the stairs, and he and Joey walked out the door back out into the rain, toward the waiting sea and their abandoned boat.

I've never seen a face look more miserable. But as Teddy later told the story, a blazing, brilliant sun came out just as he and Joey reached the boat, and they had a marvelous sail home.

Our Jewel

He was determined, dedicated, loving, and
compassionate. He was a thoughtful and considerate
person. He always had the capacity, and the desire,
to make difficult decisions. Those who loved him saw
this in him, and understood.

—ROSE FITZGERALD KENNEDY, ON BOBBY KENNEDY

I do not have a favorite."

That is how Mother answered us every time we asked which one of us was her favorite child.

"No, children. Stop asking me. I do not have a favorite. Every one of you brings your own unique quality to this family, and I love you all the same."

Bobby and Mother at the Embassy Residence (London, 1938)

We knew she was telling us the truth—almost. There was no question that we were equally loved. But we also believed that, deep down, in a small corner of her expansive heart, Mother had a soft spot for Bobby. She would never say it, but we just knew it. Dad felt the same way. And why wouldn't they?

Bobby held a special place for each of us. The seventh child,

Bobby was surrounded by girls: Pat, Eunice, Kick, and Rosemary to the north, me to the south. He longed to explore the world with Dad and to engage in debate with Joe and Jack. But when he was a toddler, the older boys were already headed into their teenage years and toward college. And Teddy did not come along until Bobby was eight years old, so he could not be his playmate for quite a while.

Instead, Bobby's daily companions were his sisters. Girls. We were headstrong, fun-loving girls, for sure. We were not overly enamored of dresses and were not afraid of getting dirty. We could stand up to anyone on a tennis court or in a touch football match. But in Bobby's young eyes, we were girls nonetheless. And for a boy, that was tough to take. Mother and Dad understood that and empathized with him.

Perhaps being encircled by females made Bobby a more sensitive, intuitive, intentional soul. Or perhaps he was simply born that way. Whatever the case, from his earliest years, Bobby was a person who could not be pinned down or easily described. In so many ways, he was a regular boy, an adventurer who was always up for a dare, who loved to find treasures in the street and secret hiding places in the yard. For a time he was a paper boy, pedaling away on his bicycle in the foggy predawn hours to deliver papers in the neighborhood. Mother was delighted by his

Bobby with baby bunny (1934)

entrepreneurial spirit, until she discovered one day that, tired of his early-morning exertion, he had convinced Dad's chauffeur to drive him along the route. Mother put a stop to that. It was not a good day for Bobby.

Bobby was purposeful, but he also had an absentminded way about him. You could not keep him out of a good climbing tree.

Bobby (Bronxville, 1934)

And if the day was a little slow and there was not much to do on the ground, look up and you would find Bobby on the roof. His knees seemed permanently skinned and bruised. He tracked grass and mud into the house, only to be turned back outside by the ever-active broom of our nurse, Kikoo. Bobby was completely of this earth.

Yet he was also somehow removed from it. Like Mother, Bobby had a strong faith, even as a boy. Sometimes he would join Mother on summer mornings for her ride to daily 6:00 a.m. Mass, where he served as an altar boy. He had a particular connection to St. Francis of Assisi, the thirteenth-century saint who gave all his worldly goods away, embraced the poor, and loved even the smallest of animals. When he received the Catholic sacrament of Confirmation, Bobby chose Francis as his middle name: Robert Francis Kennedy.

Like St. Francis, Bobby loved all animals: dogs, cats, chickens, rabbits, exotic turtles, foul-looking reptiles. The ones on four legs seemed to follow him everywhere. He doted on a large pink-and-black spotted pig named Porky. I do not recall how Porky came into Bobby's life, but for a while he was a permanent fixture. Porky even rode with Bobby to school one day, sitting up tall in the backseat of the car and looking out the window at the sights around him as they motored down the road. Imagine the people along the way, stepping onto their front porches in the crisp morning and reaching down for their newspapers, only to spy Porky and Bobby zooming by.

Bobby also gave away rabbits one year to his Bronxville friends. These children would come to our house simply to play but would leave carrying a new floppy-eared pet. No doubt the parents of the neighborhood were not thrilled when their sons and daughters arrived home with a wild rabbit thumping away in a cardboard box. "But Bobby Kennedy made me do it!"

"If you have men who will exclude any of God's creatures from the shelter of compassion and pity, you will have men who will deal likewise with their fellow men." These powerful words of St. Francis capture how my brother Bob lived his life. His love for animals was simply a microcosm of how he loved the people around him. "You have a lot on the ball and you have a personality that will make friends, so by all means hop in there and meet everybody," Dad once wrote to him. And that was what Bobby always did. Storekeepers in the village, children at school, the visitors who dropped by to see Mother and Dad—Bobby had a way with all of them.

Two of his favorite friends were the sons of a lovely woman who lived in Hyannis Port village and came to help in our house. The moment they entered the drive, Bobby was out of the house and, in a flash, the three of them were off, playing football on the lawn or pounding the basketball on the pavement outside the garage.

These were the boy playmates Bobby had longed for, and their friendship was fast and easy. Only looking back now does it occur to me how uncommon it was during that time in American history for children of different races to play together. It was never noted or even noticed in our house, or by Bobby.

When rainy days forced Bobby indoors, he explored the world through his extensive stamp collection. Ever astute, Mother recognized that Bobby was curious about other people and lands at an early age, so when he was about nine or ten, she presented him with an album to collect stamps and encouraged him in the pursuit.

Nearly every week, Mother and Dad would receive correspondence from across the nation and around the globe: postcards from friends on holiday in Europe; letters from businessmen working in Chicago and Washington, DC; parcels from shopkeepers in London and Paris. Each delivery carried on it a small treasure for Bobby: a stamp. Often they were the standard size, with profiles of bejeweled princesses, bearded kings in military uniforms, and other important people who ruled distant lands. But sometimes the stamps were larger, more colorful, and mystifying: a green upside-down triangle from Latvia; a warrior from Papua New Guinea aiming his bow and arrow; determined American workers overcoming the Great Depression; exotic birds from the Orient perched on equally exotic trees.

After removing the important contents, Mother and Dad turned each empty envelope over to Bobby. He would carefully tear off the corner that held the stamp and place it in a dish of warm water. There the stamp would sit, sometimes for hours, until its glue started to slowly dissolve and its corners began to peel up from the paper.

Bobby was anything but patient in everyday life, yet he could wait for days if he had to for his stamps. He had learned from experience that ripping a stamp off the envelope too early meant ripping the stamp in two. By placing it in water, however, and practicing patience, the stamp would eventually wrest free from its paper and float to the top of the dish. Bobby would remove it with tweezers and place it on newspaper to dry. Only once all the dampness was gone would he dare to pick it up with his fingers and paste it in his stamp album. Magnifying glass to his eye, he then examined each stamp intensely.

"Jeannikins, look! This one is about the founding of Australia. They created it for the one hundred fiftieth anniversary." "Look at this pink one, with the elephant, from Vietnam!"

Inevitably he would pull a book from the nursery shelves to find out more about Australia, Vietnam, India, Egypt, Brazil, or whatever country his stamp introduced him to. In this way, my brother Bob spent his long afternoons traversing countries and continents, while outside, the rain came down.

At the beginning of Bobby's stamp-collecting career, in the early 1930s, Dad was often in Washington, where he served as President Roosevelt's first chairman of the newly formed U.S. Securities and Exchange Commission. Those were somber times in our nation's history. America was just pulling herself out of the Great Depression set off by the stock market crash of 1929. Dad's knowledge of the financial markets was critical to getting the SEC, a revolutionary new regulatory body, off the ground.

Still, between his long, heady hours at work, Dad found a moment to mention to President Roosevelt Bobby's new interest in stamp collecting. Roosevelt was himself an avid collector, and he found a moment as well to dictate a note to Bobby encouraging him in his new hobby. Kindly, the president sent my brother some special stamps and a small album to keep them in, too.

"Perhaps sometime when you are in Washington you will come in and let me show you my collection," the president wrote.

The letter from the president arrived at Hyannis Port, stamped and addressed to Bobby. It was the source of great excitement, and Bobby immediately set out to write the president back.

"I liked the stamps you sent me very much and the little book is very useful," Bobby responded in his sprawling fourth-grade

July 12, 1935.

Dear Bob:-

Your Dad has told me that you are a
stamp collector and I thought you might like
to have these stamps to add to your collection.
I am also enclosing a little album which you
may find useful.

Perhaps sometime when you are in
Washington you will come in and let me show
you my collection.

My best wishes to you,

Very sincerely yours,

Robert Kennedy,
Hyannisport,
Massachusetts.

(Enclosure)

Letters between President Roosevelt and Bobby

Letters between President Roosevelt and Bobby

cursive. "I am just starting my collection and it would be great fun to see yours which mother says you have had for a long time."

We eventually did go to the White House, hair plastered in

Letters between President Roosevelt and Bobby

place, patent leather shoes shined, faces scrubbed. We girls wore our Sunday dresses; the boys, their suits and ties. Bobby had his stamp collection tucked tightly beneath his arm.

The president's secretary led us into the Oval Office, and

President Roosevelt, sitting in his wheelchair, greeted us there. "Which is the oldest? You are all so big!"

He asked each of us our name and was curious about what we were learning at school. Then he turned to Bob.

"Mr. President, I brought my stamp collection for you to see," Bobby announced.

President Roosevelt was delighted. The two of them moved to the large desk, where the leader of the free world brought out his stamp collection, too. There they lingered, the president and the boy, necks craned over their albums. The rest of us looked on quietly as they passed a magnifying glass between them and crossed the globe together.

Several years later, Bobby would be old enough to set off by himself to the lands he had explored through his stamps. He ventured out on a long tour of South America, discovering for the first time countries and peoples he would return to visit many years later as a U.S. senator. At one of the outdoor markets, Bobby stopped at a booth and spied what he felt was the ideal gift for me. He purchased it, wrapped it securely in tissue paper, and packed it away in his suitcase for the long trip back to America.

Once home in Bronxville, after unpacking his bags and putting his clothes away upstairs, Bobby descended to the living

Bobby, Mother, and Pat (Hyannis Port, 1942)

room, where we were all gathered to hear about his adventures. He held an enormous mound of tissue paper in his hand.

"Jean, I found the perfect present for you," Bobby declared.

I was thrilled to be singled out. I had teased him endlessly that he often brought my sisters gifts from his travels, but never one for me. Now here one was.

He handed me the tissue paper, and I began, eagerly, but carefully, to unwrap it. Layer after layer I peeled back, searching for the surprise inside. I peeled and peeled as the others looked over my shoulder. On about the twentieth layer, I finally uncovered my prize: a tiny little jewel tucked away in the crevices of the tissue. It was the smallest stone I had ever seen, no bigger than the head of a pin. It was a pale blue, I believe, but you would have needed a microscope to be sure.

I looked up at Bobby, and he grinned back at me. Bobby had this way about him: you did not know if he was putting you on or if he was dead serious. Was this gift a joke or was he truly excited? His smile seemed so sincere, so eager to please, but then there was that slight glint in his eye. I did not quite know, and I was not going to take the chance of hurting his feelings.

"Oh, it's beautiful!" I exclaimed.

"I'm glad you like it. I'm so glad. They don't have many of these down there. Isn't it remarkable?" Bobby replied. "Isn't it beautiful, Mother? Eunice? What do you think?" He called everyone over for a viewing.

They stood in a circle around me to inspect the gift. First they saw nothing, until I pointed down and they focused on the speck of a jewel. They did not know if Bobby was for real, either, but they, too, were not going to risk hurting his feelings.

"It's lovely, Bobby!" Pat said. "Just right for Jean."

"I thought you'd think so, Pat."

Then he stopped. Looking down again at the jewel, his brow furrowed, Bobby considered what he saw.

"Hmmm. It might be just a little small. Mother, maybe you can help me dress it up?"

Dress it up? What was Bobby driving at? Mother understood perfectly what he meant, and the soft spot in her heart for Bobby grew even softer. She was very taken by the fact that he had thought to bring me a present, and she wanted to make sure I wore it.

"Of course I will, Bobby," she assured him. "Let's go into town and see what the jeweler can do."

So the very next day the two of them set off, mother and son, to the jewelry shop. The bespectacled gentleman behind the glass counter excavated the jewel from its paper, lifted it with a tweezer, and placed it, with a flourish, onto a black velvet cloth. Lips pursed in concentration, he intently inspected Bobby's find through his highest-powered magnifier. Ever so seriously, he raised his head. "Mrs. Kennedy, I believe I have the perfect setting for this gem," he pronounced.

With alacrity and focus, the jeweler conveyed several gorgeous stones from his case: two substantial diamonds, two good-size sapphires, and several aquamarines. He positioned them on the velvet around Bobby's minuscule stone. "This would make an absolutely exquisite brooch," the gentleman said.

Mother quickly calculated the cost in her head. "Don't you have something slightly simpler?" she inquired.

"Oh, Mother," Bobby pleaded, sensing her hesitation. "I really do think this is perfect for Jean. These other jewels set hers off just right."

Mother looked into Bobby's face and caved. She told the man to set the jewel. And since she had gone that far already, she told him to set two.

"A brooch always looks better with a companion," she explained to Bobby.

Several weeks later, Bobby and Mother returned to the shop and collected the gift. The brooches were beautiful, Bobby's minuscule blue treasure tucked in between its glamorous larger cousins. Mother paid the man, and they returned to the house, Bobby carrying the parcel. For the second time that month, he proudly presented me with a gift.

"For you," he proclaimed.

I opened the box and inhaled at the sight. No smaller jewel had ever received such lavish treatment. Like the parable of the loaves and fishes, Bobby had taken a stone of no consequence or size and made it multiply—with an enormous assist from Mother and her bank account. He beamed, and I beamed back.

A few months on, back at the house in Bronxville, we were preparing for Christmas when a gigantic package arrived at the door addressed to me. It was from Bobby, who was still away at boarding school. Though I should have put it under the tree and waited until Christmas morning to open it, the excitement was too much. With Mother and my sisters at my side, I ripped the paper off. It was a jewelry box, the largest one I had ever seen, made of glossy black lacquer and inlaid with an intricate pattern of intertwined flowers.

I opened the lid. Resting on the creamy velvet lining inside was a handwritten note from my older brother. It was a simple message, but it made all of us wonder again: had he been putting us on the entire time?

> Jean:
> This box is to keep your jewel safe.
> Merry Christmas.
> Love, Bobby.

None of us could resist that boy.

9

On the Town
with Dad

For parents who have or expect to have a sizeable
family, my strong advice would be to work hardest on
the eldest, for in the direction they go the others
are likely to follow.

—ROSE FITZGERALD KENNEDY

*B*irds of a feather flock together. That's how the saying
goes. And that's how we Kennedys went.

Every day, we fanned out across our world. We would go on
foot, grab our bikes, or pile in the car. To the shore, to church, to
the tennis court. In twos, threes, fours, or fives, but very rarely

alone. A brother or sister was always along, urging us on, prodding us to run faster. "Come on, Jeannie, pick it up!"

Mother and Dad made sure it was so. They entrusted us to one another. It was their greatest gift to the nine of us. No matter how far away we were from one another, Mother made certain that we were always connected. When she or Dad traveled, they religiously wrote letters to each other and to every one of us, asking questions about the happenings at home. And we responded with our daily news.

Mother was a consistent and disciplined letter writer. If someone invited her for tea one day, she wrote a letter of thanks the next. If someone suffered a death in the family, she penned a letter of condolence immediately after hearing the news. Each day, after attending Mass and finishing breakfast, Mother retired to her room for her daily correspondence. It was very important to her, and she made certain that we adopted the same good habits. Each of us had stationery and pens for the job. She taught us the proper etiquette, the proper salutations, and the proper valedictions. If we failed to thank someone, we heard from Mother or Dad:

"Why didn't you drop Margaret a little note and thank her for the present she sent you, rather than for me to do it for I am sure she would love to hear from you and it would be very good practice," Dad wrote to a ten-year-old Bobby.

We even sent thank-you letters to Santa Claus.

In fact, even before we could actually write, we were "writing." In the nursery, we asked for help from our nurses, who listened as we recited our letters and took down every word in their lovely Palmer Method penmanship.

"Dear Daddy," began one of Teddy's letters to our dad when he was traveling for business.

> Everybody in the world is good but Jeannie. Jean is good sometimes too. Bobby sold some rabbits and when you come home you'll see Kikoo is home now. When are you going to telephone me again? Eddie [Moore, Dad's secretary and Teddy's godfather] took me to see the fishes, but the place wasn't open, so we fed the pigeons and they came right near me. That's all. Love . . .

The correspondence was signed at the bottom in a scratchy, four-year-old's handwriting: "teddy."

As we grew up and reached the age to go away to boarding school, Mother insisted that we call home every Sunday to report in. But making a long-distance phone call in those days was very costly, an enormous extravagance, particularly when multiple children were involved. So Mother and Dad kept the calls very short, just long enough for them to hear our voices and confirm that we were not ill or distressed. "Hello Mother, it's Jean!" I called out into the crackling receiver. "Hello, Jean, dear, how are you?"

The smaller, more minute, day-to-day details of our lives (our wins at sports, marks in class, friends we met) were saved for our letters, which we wrote weekly without fail.

"Dear Mother" or "Dear Daddy" our letters dutifully began, followed by reports from school and questions about home.

> Dear Daddy, I am sorry I could not speak to you on the
> telephone last Sunday. We had a Halloween party at school.
> It was swell. We had a hockey game at school Tuesday. We
> won and I was on the team . . . Loads of Love, Jean.

Once she finished reading each letter, Mother would make a copy of it with a mimeograph machine and share it with the rest of the family. If we were home, she would hand it to us to read. If we were away at school, we received it in the mail. In this way, Mother let each of us know how the others were faring. It was her way of making sure that we never felt too far from one another, even if we were halfway around the globe.

$\mathcal{M}$other and Dad believed that if you set your eldest children on the straight and narrow, the others would fall in line. When Teddy was born, Jack wrote Mother a letter from his school in Connecticut, asking if he could be godfather to him just as Joe was godfather to me. Mother and Dad enthusiastically agreed. In their minds, it only made sense. The older ones would feel responsible to us, and we would look up to them.

Rather than scolding one of us for bad manners, bad posture, bad grammar, or bad form, Mother enlisted a brother or sister to do so. She understood that we were much more likely to listen to our siblings than to our parents.

One day, hoping to look glamorous, I globbed cherry lipstick on my lips. Well, I was aiming for my lips, but in fact hit above and below the lips as well. I looked like a ten-year-old clown.

Mother spied me coming down the stairs. She pulled Bobby aside.

"Bobby, look at Jean," she said, making sure I could not overhear. "Perhaps you could remind her very nicely that lipstick does not go all over the face."

My brother met me in the hall.

"Going a little overboard with those lips, aren't you?" he said as he passed.

Eunice, Dad, Kick, and Pat (New York, 1940)

I immediately raced to the bathroom, found a tissue, and wiped it off.

If Mother had said it, I would have been angry. When my big brother Bobby said it, I paid attention. He represented a constituency of boys in my future who might never speak to me again if I kept wearing lipstick that way. Who knew how many of his friends might ask me out one day, as long as I didn't look like I was joining the circus.

Our responsibility to one another extended to outings as well. On special Sundays in Bronxville, when everyone was home from school for a holiday or a weekend visit, Dad would make an announcement: we were going for a drive into the city. Before piling us into two cars for the trip (we could not all squeeze into one), he would gather us together for our specific assignments:

"Joe, Jack, and Rosemary, you stick together. Kick and Eunice, you stay close. Bobby and Teddy, you sit together. Pat, keep your eye on Jean."

Then, breezing south along the Hudson River, we would alight in the shining center of it all: New York City. Broadway. Our cars would stop at Forty-Ninth Street and Madison Avenue, just down from the queen of department stores, Sax Fifth Avenue. And just

up the street was the only restaurant that most of us had ever visited: Longchamps.

With nine children, it was a very rare occasion for us to go to
a restaurant. Mother rightly believed that eating at home was a
more economical and less chaotic option. So if we ever dined out, it
was a special event. Longchamps held a magical spot in our lives
and in our dreams.

We crossed the street and filed, two by two, through the door.
The doorman nodded his head to each one of us as we passed: "Welcome to Longchamps, sir." "Welcome to Longchamps, madam."
"Thank you!" we intoned. Dad, who was just excited as we
were, brought up the rear.

The elegant interior was air-cooled and dazzling. Murals and
mirrors decorated the walls. Sliding across the banquette, we
took our seats and opened our enormous menus stamped with a
beautifully scripted *L*. The offerings were exotic and alien: cur-
ried lobster, frog's leg platter, New Orleans prawns, Blue Point
oysters, soft shell crab, baba au rhum. Our eyes gobbled up the
choices, which were so different from what was on our table each
night. What to pick?

Dad made the decision easy for us. Every time we went to
Longchamps we ate the same meal: "Roast beef and Yorkshire
pudding for the table," Dad would announce to the waiter. It was

his favorite, and we loved it, too. Delivered on steaming plates, the meat pink, the pudding light, we kept our eyes on Dad. When he raised his knife and fork, we followed suit and dove in. "Yuumm."

We hardly spoke as we finished our meals. Then, as the elegant waiters in white coats began to clear our places, Dad made his pronouncement: "It's time!"

Dad was always on time. He felt it was the ultimate insult to keep someone waiting. And he expected all of us to be on time, too.

Once, Kick scheduled a lunch date with Dad at noon in the city, but she lost track of time while shopping with friends on Fifth Avenue. When she finally arrived at the restaurant, thirty minutes late, she met him coming out the door. "The next time, be on time," he said, as he calmly walked past her. I still imagine her there, standing wide-eyed in front of the restaurant, not knowing what to do, as he walked away down the sidewalk toward his next appointment. It was a lesson that neither she, nor any of us whom she later told about it, ever forgot.

So when Dad said, "It's time!" we put our napkins on the table, slid out of the booth, and were on our way.

"Goodbye, sir. Goodbye, madam. I hope you enjoyed Longchamps."

Two by two, we walked up Forty-Ninth Street and across Fifth Avenue toward Sixth. There it was: Radio City Music Hall.

Kick

As far as we were concerned, it was the most famous building in New York. The brainchild of philanthropist and developer John D. Rockefeller, the music hall was only a few years old at that time but was already a mecca for anyone who loved the movies. And all of us adored them. To come with Dad into the city, and to see a film at Radio City Music Hall, was the ultimate treat.

We hustled forward together under the towering marquee, which seemed to glow with a million lights: Lionel Barrymore,

Evelyn Venable, and my favorite, Shirley Temple, the dimpled little girl with corkscrew curls who tap-danced her way into our imaginations and hearts. She spoke so sweetly and sang so adorably, it was hard to believe that anyone could be that cute. And in fact, in the years since, I don't believe any child star has come close to matching her box office appeal.

We saw all her films, but this Sunday, Radio City Music Hall was showing *The Little Colonel*. Shirley played a little girl living in the South during the Reconstruction who warmed the heart of her crusty grandfather, played by the great Lionel Barrymore. We had been looking forward to the film for weeks.

My sister Pat held tight to my hand as she steered me through the throbbing crowd in the Radio City Music Hall lobby. I looked up to Pat and wanted to be like her, always. In my eyes, she was like the actress Rita Hayworth, a natural beauty with auburn hair. She always looked divine. It all seemed so effortless to her.

Pat was my caretaker throughout our childhood. Four years older, she was innately kind and made sure I was okay. When we were both away at school—she at Rosemont College in Pennsylvania and I at the Convent of the Sacred Heart about thirty minutes away—Pat would plan fun outings on my days off, meeting me at the train station and taking me to lunch or out shopping with her gaggle of friends. It was such a thoughtful thing for an older, glamorous sister to do for a younger fledgling. But I also could test her patience.

"Jean, do you have to use the restroom?" Pat said, turning to me as we entered the theater at Radio City Music Hall. I tugged at her hand, ready to take my seat. The megaphone-shaped stage of the Music Hall beckoned me: *Jean. Jean. Jean.*

"No, I don't have to go," I replied.

"Are you sure, Jean, because I don't want to leave the film once it starts."

The theater darkened. We were going to miss the start. I had to move her along.

"I'm positive, Pat," I huffed anxiously. "I don't have to go!"

She accepted my assurances, and in we went.

Our family took up nearly an entire row. We settled in precisely on time, just as Dad liked it. The velvet curtains glided to either side, revealing the imposing screen, and the movie began. The South just after the Civil War. A stubborn, mean colonel named Lloyd throws his daughter out for marrying a Yankee. Years later, she must leave her husband out West, where he is working, and return home with their little girl, whom they also have named Lloyd. ("That's an unfortunate name for a girl," I

Pat

whispered to Pat. She nodded her agreement.) The angry grand-father first wants nothing to do with his granddaughter but soon begins to melt under her charms.

The story reached out and grabbed us. Little Lloyd tap-dances with her grandfather's butler up the staircase to bed. The colonel softens all the more.

Lloyd's father returns and he and the mother are soon taken hostage by two crooks after their money. The little girl runs through the dark woods to get her grandfather's help, but he refuses.

Pat's face glowed in the screen light, and her fingers gripped the arm of the seat.

But I started to twitch.

Then I started to squirm.

I tried to ignore it, but it was impossible to ignore. All those lemonades at Longchamps had started to take their toll. I had to go to the restroom. And badly.

"Pat," I whispered, just as little Lloyd stands up to her grand-father.

"Pat," I urged again, as the colonel is moved by his grand-daughter and agrees to go with her to save her parents.

"Pat, I have to go," I whimpered.

"Shhhhhhhhhhhhhhhh," she pleaded.

"Pat, I really have to go—now!" I cried.

"Oh, Jean!" she moaned. Grabbing hold of my wrist, she dragged me out the row, up the aisle, and out into the lobby. By the time I was finished and we had returned to the dark, the par-ents had been saved, Shirley Temple and Lionel Barrymore were hugging, and "The End" was scrolling across the screen.

"What happened? What happened?" Pat asked the others as we filed out of the row.

In the dark, I blushed. I had made her miss the end of the film. And I had missed it as well.

"I'm sorry, Pat." I was close to tears.

She looked down with resigned but understanding eyes.

"Don't worry," she said. "We'll come back again."

Out the front doors of the magnificent music hall and back down Forty-Ninth Street we walked, hands clasped: Pat always taking care of me and always the good sport.

A Life Full of Lessons

*I have always felt that the more experiences a child
has and the more things he sees and hears, the more
interested in life he is likely to be, and the more
interesting his own life is likely to be.*

— ROSE FITZGERALD KENNEDY

$\mathcal{M}$other was devoted to the cultivation of the mind.
And she felt that the fastest way to achieve proficiency in any pursuit was through lessons.

"Kick, have you practiced your piano?"

"Please run through your multiplication tables, Pat."

"Eunice, could you please take Jean to the tennis court and work on her stroke for a while?"

It makes me laugh to remember how often Mother and her

lessons wreaked havoc on our lives. She could not comprehend why anyone would ever pursue an activity or sport without learning the proper technique. She felt that lessons strengthened our knowledge and resolve.

Piano, golf, sailing, French, painting, ice-skating—Mother encouraged us to explore every subject and hobby that interested us, and even some that did not.

"I really think it would be a very good idea if you went to dancing school," she wrote to Bobby. "I know that you loathe [dancing lessons], but . . . I can see where practice every week would make a lot of difference in your confidence and in your dancing ability."

Dad heartily agreed with her philosophy.

"I am glad that you are sailing and doing so well," he wrote me one summer. "I think if you really got interested in it, with your good little head, you could make your sisters hustle. Also do keep after tennis because being proficient in sports helps you to get a lot more fun out of being with people."

Not surprisingly, Mother and Dad enlisted the older siblings to help the younger ones in our lessons. Some of our activities tested their patience. One year, I drove my brothers and sisters mad practicing my tap dancing on the wooden floor under the arch in the front hall in Bronxville. I wanted to be the next Shirley Temple.

Me in my Highland Fling costume (1936)

Mother informed me that I first needed to perfect the steps. It was hard and noisy work, as Mother most probably knew it would be. "What a racket!" Eunice moaned as she passed through the hall. I soon gave it up, and Shirley Temple remained solo in the spotlight.

I drew the same reaction when I took up the traditional Scottish

Mother and Eunice (1941)

dance, the Highland Fling. I can only imagine the figure I cut as I bounced through the house, in kilt and argyle socks, kicking out my legs and swinging my arms back and forth above my head.

Other pursuits were more tolerable to our older brothers and sisters. Every summer at Hyannis Port we took two swimming lessons per week with our instructor. But on top of that, Mother asked our siblings to spend time reinforcing our schooling in the water. Eunice, a natural teacher, was particularly gentle with Teddy and me, observing our crawl or backstroke and then adjusting our feet so that we moved more smoothly or faster. "If you hold your hips higher, Jeannie, you won't struggle so much," she advised me. I still enjoy swimming today thanks to my sister Eunice and my mother.

The same was true of tennis, where in addition to regular lessons, we spent hours on the court volleying, pushing one another to improve. Our constant practice made us strong tennis players, and often we ended up facing off one another across the net in the finals of local tournaments. Jack and I were usually matched up as doubles. He had a natural swing but was not as fast; I was fast with the lesser swing. Pat and Eunice were our regular opponents. We laughed a lot.

One time, in a particularly heated game, I was up to serve. Just as I tossed the ball in the air, Jack told me a joke. It was ridiculously silly, and I collapsed into laughter, missing the ball entirely. I got back in position to serve, but no matter how hard I tried, I could not pull myself together. For the rest of the game, I was useless. We lost, but I still laugh out loud thinking about it.

*M*other's love of learning all things was most certainly instilled in her by her parents. Grandpa was a largely self-taught man. He read up on absolutely every subject, and re-tained it all, too. One of Mother's favorite poems was written by a clever Boston poet, whose name is long lost. It ran in a Boston newspaper:

> *Honey Fitz can talk you blind*
> *On any subject you can find.*
> *Fish and fishing, motorboats,*
> *Railroads, streetcars, getting votes,*
> *Proper ways to open clams,*
> *How to cure existing shams;*
> *State Street, Goo-Goos, aeroplanes,*
> *Malefactors, thieving gains,*
> *Local transportation rates,*
> *How to run the nearby states;*

On all these things, and many more,
Honey Fitz is crammed with lore.

Grandpa and Grandma made sure their six children applied the same discipline in their own lives. Mother's schooling was rigorous at Dorchester High School and beyond. She spent long hours at the piano scrolling through scales and doing other exacting exercises. Her parents took her and Aunt Agnes on a two-month tour of Europe, where they walked in hushed steps under the vaulted ceilings of cathedrals and along the rambling passages of museums. The Middle Ages, the Renaissance, Shakespeare, Raphael, Da Vinci—their young minds expanded as they took it all in.

Mother and Aunt Agnes had formed a solid foundation in Latin in school and were well on their way to learning French. So at the end of their trip, their parents decided they should remain in Europe for the coming year, at a convent school in Germany, to further improve their language skills. Mother immersed herself in that quest, and had achieved a very high level of proficiency in both French and German by the time she returned home. But her pursuit of those languages did not stop there. Practice makes perfect. There were dialects to hone and diction to refine. French came more easily, but German was a particular challenge for her. For the rest of her life, well into her nineties, most often in the late evenings, we could hear a scratchy voice blasting from the phonograph in Mother's sitting room:

Kannst Du mir mehr über Dich erzählen? ("Can you tell me more about yourself?")

"Kannst Du mir mehr über Dich erzählen?" Mother would dutifully repeat.

Ich würde glücklich sein. ("I'd be happy to.")

"Ich würde glücklich sein." She would respond.

Mother even went to modeling school as an adult, to better understand how the professionals pulled off their exquisite posture and bearing. In class alongside a string of aspiring models several decades her junior, Mother discovered which colors suited her skin tone and how to pose in photographs to look her most attractive—chin up, hands on the hips, and arms never against the body. She always wanted to look smart and well dressed for Dad. In addition to using the tips herself, she made sure to pass them down to us.

"Don't wear brown, Jean. That's not flattering," she told me. "Always wear blue, because you have beautiful blue eyes."

With such discipline in her own life, it is no wonder she worked to instill it in ours.

Mother and Dad closely investigated the schools they chose for us to make sure they were up to the job of our education. When they moved from Boston to New York, they selected their new home primarily because the local Bronxville public schools had a stellar reputation and we could easily walk or bike there in the mornings. Mother felt that walking to school was good for our spirits. She also felt it was important for us to receive a public school education so we could meet and become friends with the children in our neighborhood.

At our local schools we learned the fundamentals: literature, grammar, arithmetic. Then, when we reached the age of thirteen or so, Mother and Dad selected a boarding school for us. The boys were enrolled in a number of different schools over the years, beginning with the Choate School in Connecticut. In contrast, Mother decided that we girls should attend only schools run by the Sisters of the Sacred Heart throughout our middle and secondary school years. In this way she made sure that we were following one consistent curriculum whether we were studying at Maplehurst in the Bronx, Noroton in Connecticut, or Roehampton outside London. It certainly made it easier for us to progress without having to adapt to a new approach to learning each time we moved to a new school.

Throughout our schooling, Mother carefully monitored our studies and assessed our marks. But she never felt our learning should start or end at the schoolhouse door.

Mother and Teddy (Boston, 1938)

At the table together, in addition to conversations about current events, she taught us mathematics, science, geometry, and
religion.

"We have twelve baked potatoes. If we divided them equally,
how many would each person get, Jean? How many would be left
over?"

"How many triangles can you find in this room, children? Look at that painting—do you see any triangles there?"

Her approach was often playful and never suffocating. Yet we were learning every second.

Most of all, Mother would read, read, read. For our entire childhood, she placed a premium on books, and covered us in them. She called reading "the most important instrument of knowledge," and she chose books for us carefully from lists that were put out by the Parent Teachers Association or the public library. Mother made sure the books were not just good stories, but instructional and inspirational. After lunch each day, she would send us off to our separate corners of the house to spend an hour with our books. Some of us enjoyed it more than others. Especially when the sun shone bright outside, we bemoaned having to sit indoors. Yet all of us in later life thanked Mother profusely for insisting that we read.

Mother adored history and would incorporate all the academic disciplines (art, literature, music) into our understanding of it. The Revolutionary War was a favorite time period, given that she was raised in Boston. She made a point of taking us on obligatory visits to the Paul Revere House, the modest gray-shingled structure in the North End of Boston. We filed through and peered at the open hearth in the kitchen where Revere's family sat, and at the battered tools in his silversmith shop. But we also traveled along with him on his midnight ride, past every Middlesex village and farm, thanks to Mother's constant repetition of the famous Longfellow poem.

> *Listen, my children, and you shall hear*
> *Of the midnight ride of Paul Revere,*
> *On the eighteenth of April, in Seventy-five;*
> *Hardly a man is now alive*
> *Who remembers that famous day and year.*

After dinner or during long rides in the car, she would recite verse after verse, stanza after stanza, in her high, commanding voice. The Fourth of July was not the Fourth of July without the recitation of this, our family poem. As we aged, Mother insisted that we learn the poem by heart—all thirteen long stanzas. We would stand before her in front of the fireplace, backs straight and arms to our sides.

We introduced the verse in strong clear voices:

" 'Paul Revere's Ride,' by Henry Wadsworth Longfellow."

And away we would gallop.

The drama was irresistible to our young minds and hearts. A

glimmer of light in the belfry height. Paul springs to his saddle as the second light burns. A shape in the moonlight, a bulk in the dark. We, too, heard the crowing of the cock and the barking of the dog. We, too, felt the damp of the Mystic River fog. Ride on, Paul Revere! The hurrying hoofbeats of time were the rhythm of our childhood.

"Wonderful, children, wonderful!" Mother exclaimed as she sat before us, hands keeping pace with the beat, mouthing each treasured word.

Mother introduced us to a different rhythm a little later in our upbringing: the Expert's Rhythm Drill. Typing was a skill she felt strongly that each one of us should acquire. So, true to form, she enrolled us in lessons to learn the proper method. One summer, it was Bobby's and my turn.

The Expert's Rhythm Drill was the preferred approach to typing in that day. It was repetitive, tedious, and often mind-numbing. Yet once the course was finished, we were guaranteed to be proficient.

"Now, class, please turn to the first page in your book." The instructor's voice still rings shrill in my ear. "There are three secrets to proficiency in typing: concentration, posture, and the position of the fingers on the keyboard. Look closely at this illustration and see where each finger should be placed. Now focus, sit up straight, and let's begin."

With that, Bobby and I started our first lesson.

The teacher assigned each of the students in the class to a typewriter. The machine sat ominously in the center of the desk, with the textbook propped up to the right side. Bobby settled into the seat next to mine.

First we learned to find the "home keys," A, S, D, F, J, K, L, and the colon/semicolon key. We tapped the first one tentatively, and the levers in the machine briefly swung into contortions. When they settled back into place, a perfect *a* appeared on the paper in front of us. That seemed easy enough.

Then the instructor launched us into our drills. Right hand only. Left hand only. Back and forth our hands would go. She started us slow and steady, then a little faster, keeping a solid rhythm all the way.

"Remember, backs straight, everyone. Concentration, posture, and position!" she called over the tapping of the keys.

She moved us on to typing sentences. We first typed a line that contained all the letters of the alphabet:

The quick brown fox jumps over the lazy dog.

Then she let us loose on the day's primary exercise. "Type the sentences on the next page," she commanded.

And the sentences on the next page were something to behold.

I see you are here. Do I not see that you are here? Have you been here a long time? You say that you have been here a long time. How long will you be here at this time? I hope that you will be here a long time. How did I hear that you would be here? They told me that you would be here at this time. Have you been to see them? Do you think you will go to see them? I think I will go to see them soon. When did you hear from them? Let us hear from you soon. May we not hear from you soon? Will you say when we may hear from you? When did you say that we could hear from you? If he saw them he will say so. I think he saw them when they were here. We will hear from him if he saw them . . .

Bobby threw his hands up in frustration.

"Why can't they just make up their minds that she is here, that she's gone to see them, and that they'll hear from them soon!" he grumbled in my direction, as the rest of the class tapped away.

"Don't read it. Just type it," I grumbled back, struggling through the sentences myself.

It was not our favorite class. I was bad at typing, but Bobby

was hopeless. He was also relentless. Up first thing in the morning, Bobby woke the entire house as he plugged through the
drills—*clack clack clack clack brnnnng! Clack clack clack clack
brnnnng!*

But I think the gibberish that the creators of the Expert's
Rhythm Drill tried to pass off as sentences simply kept distract-
ing him from the mission at hand.

Frowzy quacks vex, jump and blight.

"What in the world could that mean?" he barked, looking up
at me from one of the sentences.

He read another:

Pack my box with five dozen liquor jugs.

"You can say that again!" he groaned.

I wrote news of Bobby's struggles to Jack, who years before
had suffered through the Expert's Rhythm Drill himself. Clearly
amused by my description of Red Robert the Rover, the name he
used for Bobby, Jack wrote me back:

Dear Jeannie:

I was most pleased to hear from you, and am fully conscious of the honor. Particularly pleasing was the report of
your activities in the field of that marvelous machine the

typewriter which saves so much time, providing you have a couple of years to spend learning how to use it. As a matter of fact, I was particularly touched by the pitiful picture you painted of that man among men, Red Robert the Rover, and you battling your way out of bed at 8:30 to do your expert's rhythm drill . . . If you don't know what the expert's rhythm drill is, Lean Kathleen will be delighted to show you on her portable.

The Expert's Rhythm Drill tested Bobby's famously stubborn will. He could conquer nearly anything, but typing almost did him in. He eventually mastered it. Still, I suspect that for years the sound of those clacking keys and our chirping instructor made special appearances in his nightmares.

But that mattered not to Mother. Bobby had learned to type.

11

Alone with Mother

*Whenever I held a newborn babe in my arms, I used
to think what I did and what I said to him would have
an influence, not only on him, but on all whom he met,
not for a day, a month, or a year, but for time and
for eternity.*

—ROSE FITZGERALD KENNEDY

*M*other opened my eyes to the world through the eyes
of dolls.

With so many children, it would have been easy and understandable for her to lump us together for all our activities. Yet she
did nothing of the sort. She made sure that we had our special time
with her and our special time with Dad. And she applied great
purpose to determining our individual interests and nudging us

Mother, Teddy, and me skating (1939)

forward in our pursuits. With Bobby, she encouraged his fascina-tion with stamps. For Jack, it was books. Eunice was tennis.

And for me, Mother gave me the gift of dolls.

When Dad was ambassador to the Court of St. James, he sent, at Mother's urging, a letter to his fellow American diplomats around the world asking if they would kindly send two dolls, a boy and a girl, dressed in the native costume of the country where they were stationed. He told them they were for his ten-year-old daughter, Jean, "who is making a collection." Dad assured them that he would reimburse the cost of purchasing the dolls and of the postage.

Within weeks, the packages began to arrive—exotic, colorful treasures from every corner of the globe: Shanghai, Calcutta, Caracas, Moscow, Bogotá. Some were almost a foot tall. One set would be outfitted in the traditional native dress of her homeland. The next was ornately attired in gold and sequined ceremonial garments.

Every package took me to a new spot on the world map. I carefully removed each new doll from its tissue wrapping and held it up for inspection. A letter accompanying the doll included interesting facts about its dress and culture. Reading those letters years later, I find it amazing how much care these diplomats, all grown men with weighty occupations, had taken in choosing a doll for a little girl.

William C. Burdett, American consul in Rio de Janeiro, explained in his letter that he had chosen to send two dolls in traditional Bahiana costume because it is "generally considered to be the most characteristic and interesting in Brazil."

Richard Ford, American consul in Montreal, consulted his friend and well-known French-Canadian historian Emile Vaillancourt on the matter. Ford subsequently wrote to Dad that he was able to obtain from Vaillancourt "the promise of a pair of 'habitant' dolls dressed according to the period of Maria Chap-delaine." Ford continued, "Mr. Vaillancourt explains that it will take some time to dress the dolls, since they aren't available ready-made so to speak, but he assures me the job will be well done and thoroughly authentic. When completed, they will be delivered to this office for presentation to the Ambassador as a token of great esteem from the Montreal Tercentenary Committee which is even now busily preparing to celebrate the city's 300th anniversary in 1942."

Elvin Seibert, American vice-consul at the Consulate General in Bangkok, Thailand, also went to extraordinary lengths:

> As it happens, dolls are not much used by Siamese children, and there are no native ones to be found in Bangkok. The costume of Siamese children in general varies so much and is so nondescript and scanty that it was useless to attempt to supply dolls representing a small boy and girl. The Consulate General has therefore had specially made for you two ten-inch figurines representing a Siamese man and woman in the traditional costume, which is much worn up-country and is seen less often in Bangkok.

The letter explained that the five-button coat the man was wearing was called a *rajapatan* and the woman was wearing the cotton *palai* and the silk *sapai*.

And it ended:

> In case your daughter may be interested in learning some-thing more about the country from which these dolls come, there are enclosed copies of the Consulate General's general information sheet, a booklet about Siam prepared by the Consulate General some years ago, and two publications issued by the Siamese Government.

What an incredible education for a young girl. And what enor-mous generosity from these diplomats, particularly given the time at which they were writing, in the spring and summer of 1939, as war approached the world.

Dad replied to each letter we received:

> I assure you that both my daughter and I are grateful for your kindness in having the two dolls made. We greatly appreciate your taking the trouble to have everything done so well. . . .
>
> Jean will be delighted to have [the dolls], rough or not and

with or without an exotic aroma. Won't you, please, let me know what their cost was, as I did not intend in asking for them that they should be a gift.

And:

Please do not for one moment worry about the delay in securing the dolls. With all that has been happening across the world during the past months no one could find fault if the matter had been completely forgotten.

In the end, I received more than two hundred dolls for my collection, thanks to the ingenuity of Mother, the follow-through of Dad, and the graciousness of the diplomats. I first displayed my collection in my upstairs bedroom at Hyannis Port, and then we moved it to special shelves in the basement, where it remains. The thoughtfulness of those strangers from around the world left a lasting impression on me. People of different cultures, religions, governments, and beliefs had all gone out of their way to make one small child happy.

Without question, the dolls on my shelf fueled a bubbling interest in me to one day leave the house and explore the countries they represented. My older brothers and sisters were already heading out on their own, traveling together and with friends across America and throughout Europe—boarding transatlantic liners for England and, from there, fanning off to Switzerland or France. Sometimes they stayed on to study abroad. Mother and Dad cheered them on in their adventures. They felt that travel and exposure to other cultures was the best education of all.

If one of us expressed an interest in a place on the map, Mother would immediately intone, "Why don't you go?" Certainly she and Dad must have had worries about sending us out on our own. It would have been only natural. But their shared belief in the importance of travel outweighed their apprehension.

One summer, Mother came to my room and asked if I would like to accompany her to Tanglewood the following weekend. Mother often traveled by herself, enjoying a short time of solitude away from the domestic responsibilities of her daily life. So it was a singular honor when she chose one of us to come along.

"Oh, yes, thank you, Mother, I would love to go!" I replied. Though I did not quite know what I was agreeing to, I knew it was something very special.

During those years, Tanglewood was emerging as the center of beautiful music for the Boston and New York arts scene. Originally a large estate named for the tangled trees that covered its grounds, Tanglewood was bequeathed to the Boston Symphony in the 1930s as its official summer home. High in the great Berkshire

Mountains of Western Massachusetts, just between the villages of Stockbridge and Lenox, Tanglewood was a cool escape for the musicians and the many music lovers who were seeking refuge from the sticky heat of the city. It was an area that naturally drew artists with its rolling landscape and breezy climes. Herman Melville had his most productive years there at his home, Arrowhead, in nearby Pittsfield. Later still, Edith Wharton lived in the Berkshires at her home, the Mount, in Lenox.

Formalities did not matter as much at Tanglewood. The music the symphony chose was lighter fare than at the concert halls in town, and the neckties of the patrons were looser. The musicians performed in a tremendous open-air structure, affectionately called "The Shed," and concertgoers spread blankets and ate picnics on the lawn under the stars.

Mother loved attending the symphony. For one, she adored classical music. But the symphony also had played a special part in my parents' life together, starting with their early romance. When Mother and Dad first began courting, Grandpa Fitzgerald was not enthusiastic about the match. Mother was Grandpa's first child, his "miracle," the apple of his bright blue eyes. So how could any man ever be good enough for her? Like many fathers before and after him, Grandpa looked on her handsome, young gentleman caller with skepticism. Dad was a bank examiner. This

Mother (1938)

was a world Grandpa knew little about. And most concerning of all for Grandpa: Dad did not have a collar button.

Prior to World War I, gentlemen wore shirts with detachable collars which they buttoned on each morning to the neck. But after the war, a new style came into fashion. Young men began wearing shirts that had the collars already sewn onto the neck. There were no buttons needed. The older men of society, like

Honey Fitz, were appalled. Dad was among the trendsetters, and Grandpa was not pleased.

So Mother and Dad had to court quietly while Grandpa got used to the idea. One of their dates was to the Boston Symphony, where they fell in love with the music, and further in love with each other. I suspect Dad was not wearing a collar button on those outings. Still that did not matter a few years later, when, in 1914, he walked Mother down the aisle after they were wed by the Archbishop of Boston, Cardinal William O'Connell. The bride's father was proudly in attendance.

For the rest of their lives, Dad and Mother enjoyed listening to classical music together. It was always playing throughout the house when they were home. And Mother often went to hear the concerts at Tanglewood during those early first seasons. And now, out of the blue, she was asking me to go.

The Friday following Mother's invitation, we rose early for the three-hour journey to Western Massachusetts from Hyannis Port. Mother had arranged for Margaret Ambrose, our beloved cook, to make chicken sandwiches for us to eat along the way. She also packed grapes and chocolate chip cookies.

Mother got into the backseat of the car on one side, I on the other. As we motored down the drive and out toward the Bourne

Bridge, she leaned her head back against the seat and closed her eyes. It was the first quiet moment for us in a bustling, hectic summer, and we both nodded off to sleep gladly.

I do not remember the specific repertoire the Boston Symphony played while we were at Tanglewood, but I do remember Mother, her back pulling up even straighter than usual when legendary conductor Serge Koussevitzky took the stage. I remember how she turned to me after each interval. "Did you enjoy that?" she would ask. And I remember how, under her tutelage and inspired by her delight, I actually heard music for the very first time, in an entirely new way.

Back then, I did not understand why Mother had chosen me for the trip. Now I do. She knew that I was shy, perhaps a little more reserved than the others, and she wanted to spend undivided time with me alone, to find out what interested me, to make sure I was okay. It was a special treat and so thoughtful, particularly since she had eight other children who also needed her attention. And it was the start of a ritual between us, a bond. I returned several times with her to Tanglewood. And then, as I grew up, we went to the opera, ballet, and other cultural events together.

Because of Mother, I developed a love for the arts that I might have never discovered. I am forever grateful to her for that. She knew the importance of the arts in one's life and that everyone should start early enjoying painting, music, and literature. She knew what I needed, and who I might become, before I knew it myself.

12

The Dinner Discussion

*We were serious about serious things, but we liked
laughing at things that weren't, including, sometimes,
some of our own foibles. Humor is a necessary part of
wisdom; it gives perspective; it frees the spirit.*
—ROSE FITZGERALD KENNEDY

*I*f you were president, what would you do?"

So began the game we played at night over dinner. Seen in
retrospect, it seems prophetic. But at the time, it was just sport,
spurred on by the turbulent and exciting times in which we
lived. There might be an interesting tidbit in the newspaper or
a congressional race that prompted the question. Dad was deeply

connected to the Roosevelt administration leading up to World War II, speaking with and traveling to Washington often, and he would bring home news from the political front.

Unemployment was down. "If you were president, what would you do?"

A hurricane had killed hundreds. "If you were president, what would you do?"

Hitler was advancing. "If you were president, what would you do?"

The 1930s laid out a dizzying number of decisions for the president of the United States, and for almost the entire length of the decade, that president was Franklin D. Roosevelt. He inherited the Depression, he created the New Deal, he confronted conflict in Europe and looming war. His actions—along with the actions of mayors, governors, prime ministers, and dictators the world over—were served up at our dinner table along with the potatoes and gravy.

Joe and Jack devoured each morsel. They dove in with theories, solutions, remedies, and platforms. We all listened, in awe. Dad was ever alert to their opinions, encouraging them to pursue a thought, or bantering with them when he felt they were off base. As we grew older, the rest of us joined in on the conversation, with our own observations and ideas.

Joe, Dad, and Jack (1938)

Looking back, I see that the dinner table was our family hub. It was the scene of our most memorable conversations and many of our most hilarious moments. Maybe it was because dinner was the rare time when we were all sitting still, together.

It should come as little surprise that much of the planning in a house with nine children revolved around meals. We never had dinner parties. In my entire childhood, I cannot remember Mother and Dad inviting a group of friends to the Cape for a sit-down meal. Instead, they reserved meals for their children and any friends we might happen to invite along. And, with all of us at the table, with any friends, and with our three Gargan cousins in the summertime, we outnumbered a good-size dinner party nearly every night.

We were all hearty eaters, especially the boys. Mother had the job of figuring out daily what to serve, when to serve it, and where to seat everybody. She expressed enormous gratitude throughout her life for the wonderful and talented cooks who helped her through those decisions and duties, managing to keep us all healthfully fed and, at least in my case, happily plump. The cooks were so good at their art, preparing delicious meals that always seemed to match up with the seasons. Summer was everyone's favorite—the season of corn on the cob, fresh tomatoes, and, on the Fourth of July, Boston cream pie. On Fridays throughout the year, we had fish in accordance with the rules of the Catholic Church. And no matter what we ate, always to the right of each plate was the obligatory glass of milk.

Dinner was our meeting place, the place where anyone who was home migrated each day to touch base with everyone else. Our meals were full of chatter, the size of our group making anything else unlikely. But they were never noisy or out of control, since Dad could not abide a racket. Conversations could range from very serious matters to the hysterically absurd. People were always telling jokes or poking fun at one another. When I brought a school friend home for the weekend, it was a test of her fortitude if she could stand up to the kidding at dinner—and better yet, if she could give it back.

"*She* was a lot of laughs!" a brother might pronounce once my friend had left for home. No higher compliment could be paid in our house, and it made me so proud if one of my friends received it.

All the fun did not mean there were no rules to follow. To the contrary, Mother and Dad were very firm about our dinnertime rituals and procedures. We all knew to arrive in the living room at 6:30 p.m., having changed out of our swimsuits and sandy clothes, bathed, donned a fresh outfit, and combed our hair. Once assembled there, Mother would play the piano while we sang or chatted. If we had invited a new friend over for dinner, we would introduce him or her around to the others. Then, as the clock rounded 7:00 p.m., Mother would stop her playing and announce that dinner was served, and we would file into the dining room for the meal.

Per Dad's instructions, our meals started promptly at 7:15 each evening. We were never allowed to be late. Our cousin Joey Gargan later surmised that Dad's insistence on timely meals had less to do with discipline than with not angering his adored cook. "If that cook leaves," Dad pronounced one evening when Joey screeched into the dining room at 7:20, "I'm going with her."

In our younger years, the smallest of us—Bobby, Teddy, and I—ate dinner earlier in the evening, at the little table with Mother, so that she could spend time speaking with just us, hearing our childhood questions and instructing us on proper table manners.

As we grew older, she decided that we were prepared to have our meals at the same time as everyone else. However, there was still no room for us at the big table. So we sat together at a table off to the side, unless one of our older siblings was away at school or visiting friends. If that was the case, Bobby would proudly leave our table and take his seat with the others, delighted finally to be part of the grown-up discussion.

Before touching a utensil, we would all make the sign of the cross and say grace: "Bless us, oh Lord, and these thy gifts which we are about to receive, from thy bounty, through Christ our Lord Amen."

I cannot remember the moment I was taught how to hold a fork or knife. I cannot remember being taught where the bread plate or the water glass went. We learned these lessons so early in life that it seems as if we were born knowing them. We all placed our napkins in our laps. We did not reach or grab for a platter but passed it, always to the left. A few special foods could be eaten with our hands, such as corn on the cob, but we understood that everything else on the table required utensils. Except for asparagus—this vegetable sparked a lively dinnertime discussion every time we had it: could we pick up the spears, or was that rude? Mother felt that, at all costs, we should err on the side of proper manners, so she insisted we eat asparagus with our forks.

Healthy debate and disagreement were regular guests at the table. Any subject was up for grabs. Someone might report a simple happening in his day: "I saw Jimmy Fisher swimming this morning." Then the chorus would begin:

"By himself? But isn't he only seven?"

"I think so, but what's wrong with that?"

"Do you think it's safe for a child to swim alone at seven?"

"Well, I'm sure I was swimming at seven."

"If the lifeguard is there, what does it matter?"

"Maybe . . . but it doesn't seem like a wise thing to do."

"Well, not everyone would agree with you."

"How else will the boy ever learn to swim?"

"But he could drown."

"Well, when do you think someone should start swimming by himself?"

"Why don't we ask the Fishers!"

And on and on and on.

Dad might casually mention that he was surprised that Joe and Jack had been out very late with friends the night before. Shock would sweep over our brothers' faces, though they would try not to let it show. How in the world had Dad known about that? They had crept so quietly into their downstairs bedrooms, and Dad had been upstairs, asleep. Only many years later did we realize that while sitting on his second-floor balcony, reviewing his papers or placing calls to Chicago, Dad could overhear our endless conversations on the porch below. So as we shared a piece of gossip or filled the others in on our antics the night before, he was all ears. Dad got no small delight in occasionally slipping a tidbit into the dinnertime talk—"Was the party fun last night?"—and watching our eyes dart back and forth at one another in alarm.

No matter how the dinner started, it inevitably ended in politics, from a Boston ward to the British parliament. The events of those years heavy with tension overseas led to heavy considerations at tables across America. The highest goal that a boy of that time could aspire to was to become president of the United States, and my brothers were no exception. We talked about who among us would be good at the job. Joe, as the oldest, seemed the natural candidate, and as he entered college and was more and more exposed to the world around him, his interest in the idea began to grow.

"If you were president, what would you do?"

It was a lighthearted game that we never conceived might come in handy someday.

13

Teddy

*I hope when you grow up you will dedicate your life
to trying to make people happy instead of making
them miserable, as war does today.*

—JOSEPH P. KENNEDY, IN LETTER TO SON TEDDY, AGE 8

Teddy more than lived up to the hopes that my father expressed for him in a letter he posted from England, on the brink of the Second World War. Teddy was just a boy of eight then, awakening to both the kindness and the sadness that this world had in store. "Make people happy," Dad wrote to him from across the sea. It was as if Dad, recognizing a singular gift in Teddy at that early age, set him on a course. The happiness, security, and freedom of others were always Teddy's prevailing winds.

Right from the start, Teddy captured our attention and that of everyone around him. It was not lost on me as a little girl that he drew the majority of cheers when we faced off against one another in the boxing ring. Later, in London, throngs of adoring newspaper photographers trailed him as he met every animal in the zoo. And later still, in the Senate chamber, chatter instantly stopped whenever Teddy rose to his feet to speak out for the poor, the disabled, the laborers, and the disenfranchised.

Like all the boys, Teddy was influenced by Grandpa Fitzgerald and got a big kick out of him. He loved to tell the story about when Grandpa would come to visit us in Palm Beach, Florida, where we would go on vacation. Teddy would drive him over to the Breakers Hotel, where he liked to spend the morning. "He would tip the hotel desk clerk to ring the bell when guests checked in—once if they were from Massachusetts; twice if they were from Boston," Teddy recalled in his book, *True Compass*. "When the bell rang twice, up would go Grandpa, introducing himself to the strangers. 'I'm John Fitzgerald. You're from Boston, aren't you?' By the end of the day, he would have gotten himself invited to lunch and dinner and would have had the time of his life."

Teddy would roar with laughter no matter how many times he told the story.

Like Grandpa, Teddy never missed an opportunity to make someone's day brighter and bring people together. He preferred loving to hating and laughing to crying. And if the opportunity presented itself, he was always ready to sing.

I remember one evening, much later in life, when we were walking down the corridors of the John F. Kennedy Center for the Performing Arts in Washington, DC. Seeing Placido Domingo walking toward him in the hallway backstage, Teddy threw open his arms and began to sing: *"O sole mio sta 'nfronte a te!"* The great tenor joyfully spread out his arms as well and folded in with my brother in song.

It is no wonder that Teddy made friends easily with nearly everyone he encountered. He was at home with presidents, prime ministers, and titans of industry, as well as musicians, taxi driv-ers, factory workers, doormen, farmers, and artists. They each in-fluenced him. Even after the briefest conversation, he would come away with a fascinating lesson for life.

"Jean, I want you to meet this gentleman," he might say, lead-ing me over to his newest friend. "He has an incredible story."

As a young adult, Teddy forged a friendship that would mark the rest of his life. Headed to Maine one summer to visit a college roommate, Teddy stopped at the home of the acclaimed and much loved American painter, Andrew Wyeth. Andrew was the son of the artist and illustrator, N. C. Wyeth, and his own son, Jamie, was just emerging as an artist himself when Teddy came to visit. This was an extraordinary family with a legacy of talent unlike any other in America. Teddy was smitten. Jamie, and his wonderful wife, Phyllis, became our lifelong friend and inspiration.

Each summer from that point forward, Teddy would return to visit the Wyeths, first to the home in the small town of Cushing, Maine, and for years later to their homes on Southern Island, and then on Monhegan Island. My sister Pat and I often made the trip with Teddy. We would arrive early enough to go out for a sail, breezing along the rocky coast and among the small, pine-scented islands. The idea of a swim was forbidding. Even on the hottest summer days, the waters of the Atlantic Ocean were like ice. But dared on by Teddy and determined never to be called a coward, I would dive in, going into minor shock as I sank deeper into the cold. Challenge met, I would scramble back onto the deck into a sun-warmed towel, and our party would head back to shore and to a long lunch.

There at the table of our country's great artists, we listened and learned—how the masters of Europe honed their skills, how the light in Maine is perfect for painting, how the things we love around us make the best subjects, and how the arts can restore you in times of despair.

Jamie Wyeth painting of Teddy on his sailboat, the *Mya*.

Though he was flattered by Andrew Wyeth's attention, the lasting influence on Teddy's artistic life was Jamie, his great friend and inspiration. We all grew to love Jamie not only for his tremendous artistic gift, but for his very warm friendship throughout the years. He went on to paint several members of our family, including what I believe is the most poignant portrait of Jack. He captured him as I remember him, casual yet engaged, leaning to his right, hand to chin, as if he were listening intently and forming an opinion about a point someone was making off canvas.

Jamie was always encouraging Teddy to give painting a serious try, however campaigns and rallies and Senate hearings always seemed to get in the way. Then, in 1964, fate played its hand when Teddy was gravely injured in a plane crash. He broke his back, and for the next six months he was confined to a specialized hospital apparatus, called a Stryker frame, while he healed. The Stryker frame is a large, imposing, metal device. It clamped down on both sides of Teddy like a toaster and held him straight as a board and entirely motionless in the center. So that his back would heal properly, the nurse rotated the frame every few hours, back and forth—upside down, where he would hang facing the floor; or right side up, where he would lie staring at the ceiling.

The Stryker frame was effective, yet painful and undoubtedly boring. But that did not vanquish Teddy. He was constantly upbeat throughout his recovery. "Mother always wants the best for her youngest!" he wrote on a photograph he sent to me of him being turned over in the Stryker frame by a burly male nurse.

Turning adversity into opportunity, Teddy realized he now had the time to pursue his dream of being an artist, so he asked us to purchase him a set of paints. While hanging upside down, facing the floor, he painted for hours on canvases positioned beneath him.

Teddy remembered what Jamie had told him, and he painted what he loved: the house at the Cape, sailboats, the inviting sea. Painting became a passion for Teddy, one he pursued long after he was released from the Stryker frame and resumed his regular life. Over time, he became very good. Teddy loved nothing more than to share his latest works with us at Christmastime each year.

"Always gentle seas and warm suns at the Cape," he wrote under the lovely painting of the house at Hyannis Port that he gave to Steve and me. It hangs on my wall still, a daily reminder of the generous, positive, and happy spirit who created it.

14

Daily Walks

*Young children need fresh air, exercise, and activities
that stimulate their interests, some of which I could
supply by taking them for a daily walk.*

—ROSE FITZGERALD KENNEDY

$\mathcal{M}$other was a great one for a walk. It was tremendous exercise. That she knew. But it was also a way to spend time— with her thoughts, with the person beside her, with the world around her. And it was definitely more fun than sitting around on the couch.

In the evenings, she and Dad often took walks around the neighborhood in Bronxville. My brothers, sisters, and I would run down the driveway to catch up with them, but inevitably

Mother and Dad headed out for a walk (Bronxville, 1938)

we would fall behind, lose interest, and go off to play. And that was how it should have been. Because it was clear that this was Mother and Dad's time to be with each other.

On summer days, Mother would invite us to join her for a walk along the shore or on the golf course. Nearly every afternoon at 3:30 or 4:00 in the summer, after her nap upstairs, she would come downstairs and head out for her daily game of golf. My sisters and I knew we were welcome, and often went along, grabbing a set of clubs that we shared from the closet. Off we went with Mother on the drive to the local course.

Mother was a steady golfer. She hit her balls straight down the middle with remarkable consistency. Her shots were short and always straight. We would be respectfully quiet of each other as we teed off, but between shots and between holes, we would walk and talk. She would ask us about ordinary things: friends, school, and the books we were reading. If we had been holding off telling her about trouble with our lessons, that we had failed an algebra or history test, this was our time to confess.

"Mother, I need to tell you something. That exam on the Roman Empire didn't go as well as I'd hoped . . ."

She would look straight ahead, intently, focused on her ball, but also on our words as we murmured explanations, complaints, excuses, and apologies.

Then came the questions:

"Did you study hard enough, Jean? What parts of the test were difficult?"

Followed by advice:

"You must never give up, Jean. You have to keep at it."

And a dose of encouragement:

"It will be fine, dear. It just takes a little more time. Don't rush too much or think about it. Just relax, and it will come to you."

The conversation completed, we would move to the next hole, her eyes still on the ball.

Mother was a common figure walking along in Hyannis Port, wearing a fashionable but comfortable outfit, always with long sleeves, often topped off by a broad-brimmed hat to keep off the sun and sunglasses to keep out the glare. Mother had a steadfast aversion to intense sunlight. She knew it was bad for a person's skin. "Put something on!" she would demand if she happened upon us on the lawn, bare arms and legs stretched out to the skies. "You'll ruin your skin. Stay out of the sun." She was way ahead of her time on that point.

Mother was not a great one for the car, and turned to it only when absolutely necessary. She did not like to drive herself, but instead would wake one of us before dawn to take her to daily Mass, or call us in from the beach to motor her to an appointment.

One year, we had what we thought was a brilliant idea: to give her a car for her birthday, so that she would not have to rely on us for transportation—and so we could sleep in or keep playing tennis as we pleased.

We pooled our money together and, with an assist from Dad, purchased a small blue coupe. We presented it to her in the circular driveway at the front of the house on the morning of July 22. "Surprise, Mother! Happy Birthday!"

Mother was, as always, appreciative, marveling at the shiny paint, and the knobs and levers inside. The next morning, she rose early as always and drove herself for the first time to Mass at St. Francis Xavier, as the rest of us, stirred momentarily by the running engine, turned back into our pillows. This was to be our new life: Mother, independent, on the road, leaving us, unbothered, with our games, friends, and follies.

But Mother was no fool. That afternoon, after we had all finished our lunches and were ready to head out again, Mother asked for our attention.

"Children, I want to thank you for this wonderful gift. It is extraordinary, and you were so lovely to think of me," she began very sweetly. "However, I do want to make just one small thing clear. I will drive myself to Mass in the morning. That is fine. But I know you still understand that I expect to be driven as always, by you, to my doctors' appointments, to the drugstore, to the shoe store, to the beauty parlor, to the golf club, and on various errands like that. You will continue to take me as always. Do you understand?"

She smiled.

"Yes, Mother," we all responded.

Yet even if she had all nine of us at her disposal to drive her, Mother's preferred mode of transportation was always her own feet. Wherever she was in the world, she made sure to get her daily walk in—along the boulevards of Paris, through the lanes in England, down the beach in Florida. Mother's walks led her in very interesting and amusing directions. Once, when her car broke down in Palm Beach and she could not contact one of us by phone, she resorted to hitchhiking, much to the amazement of the couple who picked her up.

Later in life, as Jack rose in politics, our once little-known family became a curiosity for tourists visiting the Cape. Buses would pull into the driveway and park, sometimes at inopportune moments or late in the night. We asked the bus companies if they could try to keep to the road, so that we could enjoy some privacy and exit the drive when we needed to, and they graciously complied.

But one summer afternoon, while Eunice, Pat, and I were eating lunch by the window in the front dining room, we looked out to see, coming around our circular driveway, a giant passenger-laden bus. We could not figure out what was happening, since, at our request, they usually stopped at the top of the street. However, into the driveway this bus rolled, right up to the front door.

We rose from our tuna fish sandwiches ready to see what was

the matter. But the scene outside the big picture window brought us to an abrupt stop.

The doors of the bus opened, and a familiar small figure stepped out. And though her face was hidden by her hat, there was no mistaking who it was.

"Is that Mother?" Eunice asked incredulously, though she already knew the answer.

Mother was smiling broadly and chatting, addressing the people who were behind her inside the bus but hidden from our view. Then, with a huge wave of her arm, she beckoned the group forward and directed them to follow her into the house.

One by one, the tourists came down the steps of the bus and marched behind her up the sidewalk and onto the porch as we came out the front door.

Mother immediately took note of our puzzled expressions and met them with a matter-of-fact response: "I was out on my walk and these nice people on the bus stopped me, looking for the Kennedy house," she explained. "So rather than reciting the directions, I got on board and showed them the way. They want to see the house very much, so I'm going to show them around."

With that, Mother walked right past us into the house, with the group quickly following in her steps. "This is the sofa where the Pope sat!" we heard her exclaim through the open door. Mother was delighted, and the tourists were, too.

Years later, while she was making one of her regular visits to New York City, Mother came to have dinner with my children, my husband, Steve, and me. Mother always loved to spend time with the children when she was in the city. She would sit around our little kitchen table and ask them interesting questions about what they were studying or about items in the news, just as she had done with us when we were young.

"Would you go to the moon if you ever had the chance?"

"If so, what would you do there?"

"What do you think it looks like?"

After dinner, as the children headed off to bed, it was time for Mother to return to the apartment that she and Dad kept in the city. It was just around twilight, and the sun was going down.

"I'll call a taxi for you," I said and started for the phone.

Mother stopped me. "No dear," she said. "I'd rather walk."

I thought it was a bad idea. It was getting dark, and Mother was aging. I pressed her further to take a taxi, but she would have none of it. I knew arguing the point was futile, so I told her I would join her.

We headed out into the cool air and strolled along Sixty-Seventh Street, toward Fifth Avenue. As we rounded the corner she said, "Let's go through the park."

I asked her not to. "It's too late, Mother, and it could be dangerous. Let's stay on the sidewalk."

But she insisted: "Jean, it's a lovely night. And I have read that walking on the grass is much better for your feet than walking on the concrete. Let's go into the park."

We entered Central Park and started down the path.

Mother walked upright and with a quick step, a black handbag with a short handle always on her arm, like Queen Elizabeth. For such a small person, she had tremendous carriage. Her body language commanded attention and respect.

It was much quieter in the park than on the street. These were the days before the city had made its significant upgrades to Central Park to make sure it was well lit and safe. Darkness gathered beneath the overhanging trees. A sudden breeze set off a chill.

Up ahead, at the other end of the park, two figures appeared on the path. They were shadows at first, but as they moved slowly toward us, their features began to emerge. Young men, no older than teenagers, in dark jackets. Frowning. Hands stuffed down in the pockets of their jeans.

I immediately went cold. I am sure Mother did, too, but she kept on walking, steadily, assuredly, head forward, as if nothing were wrong.

I breathed in and hoped we would pass the boys without incident. But as we neared, one of the teenagers stepped in front of us. He was large and surly. And I knew instinctively he was trouble. I closed my eyes for a second, hoping he would go away. I opened my eyes.

He was still there.

He held a cigarette up to my mother. "Gimme a light, lady." It was a demand, not a question.

I sensed movement out of the corner of my eye. The hand that was not holding the cigarette was going for Mother's handbag. It was all happening too fast. I needed to run, but how could I run with Mother. What if they knocked her out? Should I scream? I was pitiful in my panic. But it turns out that Mother was completely in control.

It did not matter that she was half his size. Back straight, she didn't miss a beat.

"Young man," Mother replied sternly. "I do not smoke. And *you* shouldn't, either." The formerly menacing young men, dumbfounded, deflated like a pair of toy balloons.

I was just as amazed as they were. I had no idea Mother knew what was going on.

And in that moment, she made her move.

"Come on dear," she said, grabbing my arm.

She marched me past the still-stunned teenagers, up through the exit at the other end of the path, and toward the street.

We emerged onto the sidewalk, so glad to be greeted by the sounds of the city, and Mother's pace returned to normal.

She stopped and turned to me. "I'm sorry, dear. I owe you an apology. You were right. It was not a good time to go in the park."

Arm in arm we walked, under the streetlights, toward her home.

15

Forever Changed

From faith, and through it, we come to a new
understanding of ourselves and all the world about
us. It puts everything into a spiritual focus . . . so
that love, and joy, and happiness, along with worry,
sorrow, and loss, become a part of a large picture
which extends far beyond time and space.

—ROSE FITZGERALD KENNEDY

$\mathcal{A}$s with any family, our together days came to an end.
In 1938, President Roosevelt appointed Dad ambassador to the
Court of St. James, elevating our family to a level of renown we
had not known before. Within days, or so it seemed, Dad sailed
to England to take up his duties. Just a month later, the rest of us
followed.

Kick, Dad, Mother, Pat, me, Bobby, and Teddy in front (London, 1938)

We traveled with Mother on the SS *Washington*. Only Joe and Jack stayed behind, to continue their studies at Harvard. Friends hailed us farewell on one side of the Atlantic, and the press greeted us with cameras on the other. We were sailing into a new era, not only for our family but for the world. War was

Teddy and me at the changing of the guard (London, 1939)

just a year away. But for my ten-year-old self, life was nothing but thrilling.

Once in London, we moved into the massive six-story Embassy Residence, which had an elevator and, to our tremendous surprise, a television. It was the first time I had ever seen one!

Soon after arriving, I enrolled in my new school, the Convent of the Sacred Heart in Roehampton, where I became familiar with the Sisters of the Sacred Heart, the same order of nuns that would teach me for years to come at other Sacred Heart schools in the United States.

At Roehampton, I boarded for the first time in my life. Eunice and Pat boarded there, too, while the others settled into the London home. The rigor of studies under the Sisters was an abrupt change from the Bronxville Public School. I spent hours reading British novels in the library at Roehampton, entranced by the longings of Jane Eyre and the escapades of Oliver Twist.

When I visited London, it was exciting. Bobby, Teddy, and I went to see the changing of the guard at Buckingham Palace and the elephants at the London Zoo.

Almost from the minute she stepped off the boat, Kick became the toast of London, so full of life and fabulous looking. She and Rosemary made their social debut in the months after we arrived by being presented at Buckingham Palace, along with Mother.

There was some whispered concern among staff that Rosemary might not be able to handle the event, but Mother and Dad insisted that she could do it, and they were right—she was sparkling in her white gown lined with silver ribbon, descending the Embassy stairs on her way to the palace. Another momentous

Kick and Rosemary, with Mother, at their presentation (London, 1938)

occasion at our house was when King George and Queen Mary came for dinner. Mother filled the house with flowers and set the long table in the dining room with the embassy china and crystal. We ate strawberry shortcake for dessert and then the group retired to watch the film *Goodbye, Mr. Chips*—with the King and Queen! It's no wonder that to me, our days in England were like a fairy tale.

While in Europe, our family took several trips, most notably to Rome, where we had an audience with Pope Pius XII. This

Our family after being received by Pope Pius XII at the Vatican (1939)

was the same gentle man who had visited us as a cardinal and sat on our sofa years before at the home in Bronxville. Dad arranged through Count Enrico Galeazzi, the acting governor of Vatican City, for Teddy to make his First Holy Communion presided over by the Holy Father. I think it was the most exciting day in Teddy's entire life.

Time would soon tell that all was not as smooth or light as it seemed to me then. War gradually made itself known in our lives. In September 1939, England declared war on Germany. Under the imminent threat of bombing raids, Mother and Dad had to decide what to do. Dad felt he needed to stay in England, and thought that Rosemary should stay with him, because she was doing so well at a school there, better than she ever had before. He was devoted to Rosemary and very happy to keep her close to him. The rest of us returned home on an ocean liner bound for the States. The ship's captain rightly feared that German planes would bomb us if they spotted even a speck of light, so he ordered a complete blackout for the length of the journey. We could move around the ship, but were very subdued. We slept together in one stateroom, completely in the dark. The ship's crew had drained the enormous swimming pool and laid mattresses on the bottom to create more sleeping space for passengers eager to leave London. Bobby slept in the pool, along with other boys his age making the voyage.

Our lives had forever changed. It was very hard to be separated from Dad and Rosemary, and they missed us as well. Dad was working incessantly, which may have made his loneliness all the worse. "Having to live this life with the family in America is nothing short of hell," he wrote. "They were with me so much the last year and a half and . . . I had such a great time with them."

To shorten the distance between us, those of us in America recorded ourselves singing and sent the records overseas to Dad and Rosemary so that they could hear our voices. For Dad's fifty-second birthday, in 1940, we belted out the Cole Porter hit "You're the Top."

You're the top!
You're the Coliseum.
You're the top!
You're the Louvre Museum.

He reported back in a letter to Mother that he absolutely loved the record and had played it "at least twenty times already."

"I can't say that any of my children's voices have improved in tone quality since I heard them last, but there is still plenty of pep in the sound. And incidentally your piano touch was never better."

Later he wrote to Mother, "Well Darling, I've dictated the news but I want you to know that I love you and miss you terribly. The excitement of this life of course keeps one going . . . I just wish I could be with you and help with the children."

Eventually Dad made the difficult decision to take Rosemary out of her school and send her home. Although she was initially happy remaining in England, her spirits began to drop profoundly from being away from us. Dad could not stand seeing her so sad. Soon after she left, he followed. The conflict in Europe continued to escalate. We later learned that our convent school had been bombed in the raids. The United States entered the war, and Joe and Jack along with it. They both joined the U.S. Navy, Joe as a pilot and Jack at sea.

Fate determined that our family would never be whole again when the news arrived, one hot August day in 1944, that Joe had been lost. Joe had written us a letter telling us that he was on his way home, and our house was full of excitement. But then he sent us another letter, saying that he had actually delayed his return by a few days so that he could volunteer for one more assignment. It was a fearless, selfless decision with tragic consequences. Courageous, brave Joe was flying that final mission over Europe when his plane exploded.

Joe

Mother and Dad learned the news of Joe's death when two local priests from the church in Hyannis Port arrived at the door that Sunday afternoon. We had just finished lunch on the wide front porch. Dad was taking a nap upstairs; Mother was reading in her sitting room. A group of us was in the living room listening to the radio. The popular wartime song "I'll Be Seeing You" had just begun to play.

Mother heard the knock on the door and opened it to the priests. They asked if they could have a moment to speak with Mr. Kennedy. Mother recalled later that she was not alarmed by their visit at first because the clergy often came seeking donations for charitable causes. She asked if Dad could speak with them later, as he had just gone up for a nap. They paused and then told her the reason for their visit—that Joe was missing in action.

She ran up the stairs to get Dad. The rest of us immediately sensed her fear. When Dad came down, he and Mother led the priests to a side room and closed the door. When they emerged, it was clear to all of us that they were stricken. Softly they told us that Joe was missing and presumed dead. It was impossible to believe. As the priests provided the details, the reality of the situation sank in. "I want you all to be particularly good to your Mother," Dad said, voice shaking.

I did not want to believe it. I stood up from where I sat and dashed out the front door, jumped onto my bike, and pedaled down to the church. There I went to the pew in the front, where I cried and prayed by myself. I then went down to Hyannis Hospital, where I was a volunteer, and began to work. What else could I do? My adored brother and godfather was gone.

Mother and Dad sustained us in those days with their faith and perseverance. They charted our course back toward joy and hope.

"When young Joe was killed, my faith, even though I am a Catholic[,] did not seem strong enough to make me understand that . . . he had won his eternal reward," Dad wrote a friend. "Rose, on the other hand, with her supreme faith, has just gone on and prayed for him."

Later Mother recalled Dad telling her, "We've got to carry on. We must take care of the living. There is a lot of work to be done."

Their courage and strength are still difficult to comprehend. Carry on they did, and we along with them.

16

A Long Way
from Bronxville

*The measure of a man's success in life is not the money
he's made. It's the kind of family he has raised.*

—JOSEPH P. KENNEDY

$\mathcal{O}$ne by one, we left the house to go to college, enter the
military, work, and pursue futures of our own. I was honored to
be chosen in 1945 to christen the USS *Joseph P. Kennedy Jr.*, a
naval ship named for my godfather, Joe. Soon afterward, I graduated from the Convent of the Sacred Heart in Noroton, Connecticut, and told Mother I wanted to follow in Eunice's footsteps and
attend college at Stanford University in California. But Mother
was reluctant to allow me to go. She felt it was very important

Me christening the USS *Joseph P. Kennedy Jr.* (Quincy, Massachusetts, 1945)

to continue my Catholic education for at least one more year. We struck a deal: I would stay on with the Sisters of the Sacred Heart at Manhattanville College of the Sacred Heart, located in Harlem, New York, for a year. If I still wanted to go to California after that year, I could transfer. It seemed fair enough to me.

On my first day of college, I walked into my dorm room and met my roommate, Ethel Skakel. A friend of mine from board-

Ethel and Bobby (Hyannis Port)

ing school had known Ethel and recommended that we room to-gether. My friend felt we would get along like a house on fire, and how right she was. It was clear from the first second that Ethel and I would be fast friends.

Ethel grasped my hand and shook it enthusiastically. "Great to meet you!" She also came from a big Catholic family, of seven children, in Greenwich, Connecticut, so we had much, perhaps

Steve and me at our wedding (New York City, May, 1956)

too much, in common. As the school year progressed and Mother received my marks, she became concerned that our friendship was taking its toll on my education. She wrote to the head of the convent asking that a makeshift wall be put up between Ethel and me, so that we could not talk all night. It was a futile effort. We simply talked over the wall.

Ethel and I arranged our schedules just right so that on Friday afternoons both our classes ended at noon. That gave us

From left: Stephen Jr., Amanda, Steve, me, Kym, and William

the time we needed to take off and head to her house for the weekend, where I got to know her parents and brothers and sisters. I also took Ethel to Hyannis Port to meet mine. I introduced her to my brother Bobby, and the rest is history. So Mother was right again. I never did make it to Stanford, and I was not sorry for it.

Bobby and Ethel were the first to marry, followed by Eunice. She found the perfect match in Sargent Shriver, who went on to found the Peace Corps and whom we all loved. Jack married next, followed by Pat. And then luck played its hand for me in the mid-1950s when I met a dashing young man, Steve Smith, just recently returned from service in the Korean War. We married in 1956 and had four children: Stephen, William, Amanda, and Kym. Steve ended up playing an instrumental role in both Jack's and Bobby's campaigns.

As my brothers grew into men, they became more and more interested in politics. Those early conversations at the dinner table became the daily occupations of their minds. Jack, always a lover of literature and words, had originally thought he would be a journalist, having written, among other pieces, the widely acclaimed book examining the British role leading up to World War II, *Why England Slept*. Yet after we lost Joe, he began to gravitate more and more toward the political sphere.

In 1947, Jack became the first of our family to run for office. He entered and won the race for Congress, representing the Eleventh District in Massachusetts. Six years later, he launched a state-wide effort, vying for the U.S. Senate against the powerful Henry Cabot Lodge, a descendant of a Massachusetts political dynasty. Naturally, the rest of us fell in line to help.

In the campaign, like on the sea, we each had our role and re-sponsibility. It was as if we had been in training for this race all our lives without realizing it. Our parents had taught us to stand up for what we believed, to support one another at every turn, to explore the world around us, and to love our country. Dad whole-heartedly agreed with Jack's decision to choose Bobby to manage the campaign. My mother, my sisters, Ethel, and I attended tea parties that ladies in towns across Massachusetts hosted to get out the vote. Sometimes we would go together, but most often we split apart: Eunice and Pat off to one corner of the state, Ethel and I to another.

Visiting the households of Massachusetts was an incredible experience. We met women of every ethnicity and interest. We talked about our brother, about our days growing up, about his heroism at sea during World War II on the small torpedo boat PT-109. Jack was the commander of the boat, which was rammed by a Japanese destroyer in the middle of the night in August 1943. Two crewmen were lost, ten others were in distress, some were terribly injured. Our wonderful older brother set about rescu-ing his men, towing them in to shore and helping them leave the wreckage and find safety.

Jack spent the nights swimming between islands looking for respite and help. He and his crew were finally saved when a native islander agreed to carry a message through the Japanese line to the American troops on the other side. Jack gave the man a coconut and scratched into its side 11 ALIVE. NEED SMALL BOAT . . . KENNEDY. It was a treacherous mission, a tremendous risk, but the man made it.

The imprint this story left on those groups of women at the tea parties was fast and permanent. They marveled at Jack's heroism and pressed us to convey their support directly to him. They were all so enthusiastic, so kind and supportive. The ladies asked us questions about our families, our children, and our home in Hyannis Port, and we answered each one. Eventually the conversation veered toward more serious subjects: domestic policies and the political differences between the candidates. When they began to question us about the state budget, that was when we knew it was time to take our leave.

"Oh, we're terribly sorry, they're waiting for us at the next house. Thank you so much!" we said as we moved out the door. "And vote JFK!"

Upon his defeat, Lodge grudgingly observed, "It was those damn teas that killed me."

It was all so much fun, and exciting. In fact, working on a campaign may be the most fun a young person can have. Over the years, as Jack continued to advance in politics, and Bobby and Teddy, too, we branched out from Massachusetts and across America, to Minnesota, Florida, Oregon. We met fascinating people, heard about their lives and their concerns, and visited remote parts of the country that we had only read about in books. Commentators and historians later said that our family's efforts made a difference in my brothers' campaigns, and I am certainly glad they did, since we enjoyed doing it so much. Looking back, I see that there was nothing more energizing or more important than playing a part in who would run the government, and I urge young people to do it today. Don't just complain. Make your voices and your opinions heard.

One time, Bobby's campaign asked me to make an appearance for him in Oregon. I boarded a plane for the West Coast. The trip was very rough. About twenty-five minutes outside the airport, the pilot's voice came over the loudspeaker. I remember his words well.

"I have very good news. We are coming in, and it looks like mostly smooth sailing from here," he said. "But on the way down"—and here he paused for a moment— "you may see whitecaps in your coffee."

I looked up from the magazine I was reading. Did that pilot just say what I thought he did? This was not comforting news, to say the least. The last thing you want to hear if you are sitting on a landing plane is that you might see whitecaps, in a cup of coffee no less. I swirled the spoon in my coffee cup faster and faster until the cream rose up like waves. If that's the type of twisting we were in for, I wanted no part of it!

The plane did indeed toss on the clouds on the way down, but in the end, we landed safely and all was well. With my feet firmly on the ground and the tension past, it immediately occurred to me that the pilot's phrase, "whitecaps in your coffee," would make the perfect title for a book about campaigns. You can be going along smoothly and then, suddenly, your opponent gets a great endorsement, or your staffer makes a mistake, or your speech falls flat, and your fortunes turn to dust.

Undoubtedly my brothers faced whitecaps along the way in their campaigns. But no matter the case, Mother and Dad never lost faith in them. Dad particularly had a way about him that buoyed you up even when you were feeling very low. At the 1956 Democratic Convention in Chicago, Jack made a fierce run to be nominated as the vice-presidential candidate on the ticket with presidential nominee Adlai Stevenson. The former governor of Illinois, Stevenson was making his second bid for the White House, and the nomination would have immediately catapulted Jack to the national stage. But the delegates chose Senator Estes Kefauver from Tennessee as the nominee instead.

Jack was despondent. We all were. He, Bobby, and I began to walk out of the stadium on the South Side of Chicago and, passing a phone booth, Bobby had an idea: "Let's call Dad."

Jack entered the booth and dialed Hyannis Port, where Mother and Dad were watching the convention at home. Dad picked up the phone, and though the door to the phone booth was closed, we all heard his booming greeting:

"Congratulations!"

We thought maybe he didn't realize what had happened. But he most certainly did.

"Congratulations, Jack," Dad continued. "Stevenson is not going anywhere. This frees you up to do what you want to do and chart your own course. This is the best thing that's ever happened to you!"

Jack came out of the booth with an enormous smile on his face. Dad had turned the entire thing around. We all began to laugh and celebrate. Dad was right again.

In the late 1950s, Jack began thinking seriously about whether he would run for president of the United States. This was a question the family had discussed back and forth loosely for years. And it was a decision that Jack asked all of us to be involved in.

Among Catholics involved in politics, the candidacy of Al

Smith, though some twenty years before, was still fresh in our minds. The first Catholic to receive the nomination of a major political party for president, Smith was pilloried by religious leaders, particularly in the southern United States, for his "anti-American" leanings. It was said that, if elected, he would answer to the Pope. There was even talk that the Pope would move to the United States. No one was surprised when Smith was soundly defeated by Herbert Hoover at the ballot box. So it was understandable that we were wary of Jack facing the same scorn.

After much deliberation, Jack finally declared his candidacy. And in fact there was a similar outcry, and he faced the same suspicion that Smith did because of his Catholic faith. It was not until September 1960, just a few months before the election, that Jack put the fervor to rest in an address before a group of Protestant ministers in Houston, Texas.

"If this election is decided on the basis that 40 million Americans lost their chance of being president on the day they were baptized, then it is the whole nation that will be the loser—in the eyes of Catholics and non-Catholics around the world, in the eyes of history, and in the eyes of our own people," Jack said.

It was a great speech, very moving for all of us, and it gave us a good understanding of the bias. In my memory, it was one of his most defining and courageous moments.

Campaigning for Jack (1950s)

The race was intense. My sisters and I took our tea parties back on the road, once again fanning out across the United States. On the night of the election, we still had no idea if Jack or the Republican candidate, Richard Nixon, would win. We spent the night moving between "The Big House" in Hyannis Port, where we spent summers growing up and where Mother and Dad now lived year-round, and a house nearby that Bobby and Ethel shared with their nine children. Members of the media, campaign workers, family members—all of us were waiting to hear the winner announced, but the numbers were too close.

Our family had dinner together, and eventually we went to bed, with everyone nervous. At dawn, we awoke to have our fears allayed. Jack had won. There was much excitement. An official photographer took a picture of all of us standing before the fireplace. Jackie borrowed my coat to go down the driveway with Jack to speak with the press. Meantime, relieved and needing to move about, the rest of us started a touch football game on the lawn. When Jackie and Jack came back up the drive, he joined in.

After a while we heard Dad shout from the porch, as was his routine, "Lunch is ready!"

We broke into little groups and began walking up toward the house. Jack and I paired up and were walking a little bit behind the others.

Ever punctual, Dad went inside only to come back a few seconds later.

"Hurry up now, everybody's ready and inside waiting for you two!"

Our family the day Jack won the election.
Standing from left: Ethel, Steve, me, Jack, Bobby, Pat, Sargent Shriver,
Joan Kennedy, Peter Lawford; Seated from left: Eunice, Mother, Dad,
Jackie, and Teddy (Hyannis Port, November 9, 1960)

Jack turned to me with a mock seriousness and said, "Doesn't he know I'm president of the United States?"

He had such a grin on his face that we both burst out laughing.

—◆—

Jack and Steve at the White House (Washington, DC, 1963)

$\mathcal{I}$n her memoir, Mother wrote about being on the campaign trail with my brothers and listening to them speak before a large audience or take on a foe in a debate. She said she often thought back to those early days around the table, when we were encouraged to learn about the issues of the world and to defend our positions. I think, too, of the Beatitudes, Jesus's lessons on love and service, which she and Dad pressed so firmly on our consciousness.

"If children are to develop into effective people, the process must begin when they are small," Mother once wrote, remembering our childhood.

I am so grateful to our parents for being so focused in their child rearing, though little could they have imagined the life they were destined to lead when they first started out together in their tidy house on Beals Street in Brookline. Little could they have imagined that Dad would be appointed the first Irish American ambassador to the Court of St. James, or that, even later still, one of their children would become the first Irish-Catholic president of the United States.

"It's a long way from Beals Street," Dad said to Mother in 1938, after they were received by Queen Elizabeth at Buckingham Palace.

A quarter of a century later, Jack was in the White House. One day, he called me on the telephone. The president of Ireland, Sean LeMass, was coming for a state dinner. The First Lady, my dear friend Jackie, was in Europe visiting her sister, Lee. Jack asked if I would serve as the hostess for the event. Would I ever!

I excitedly began to prepare. I looked at photos of my elegant sister-in-law at other state dinners, to determine what to wear.

Irish President Sean Lemass, Mrs. Kathleen Lemass,
me, and Jack at the State Dinner (Washington, DC, 1963)

Jackie was the essence of grace and style. Brilliant, beautiful,
with a quick humor and enormous heart, she never put on airs
or became self-important, even though she was arguably the most

famous woman in the world. The idea of filling her shoes at a state dinner was daunting to say the least.

I searched the stores and finally decided upon a lovely light blue gown with a lace collar. But what to do for the tiara that Jackie always placed so delicately in her hair? I did not have such a thing. So I searched my jewelry box and found a rhinestone bracelet and tucked it into my hair, perched precariously above my forehead, secured with two bobby pins. This would have to do the job.

I arrived at the White House and was escorted to the top of the stairs, where Jack waited for me. "You look great," he said, taking my arm. I smiled back at him, making sure to keep my head perfectly straight so that the bracelet would not go crashing to the floor.

We turned and stood to face the crowd below. Mr. and Mrs. LeMass looked up at us from the base of the stairs. The guard announced, "Ladies and Gentlemen, the president of the United States."

We slowly began to descend. When we reached the bottom, on the last step, Jack paused and smiled at me.

"It's a long way from Bronxville," he said.

I turned my head up to him, very carefully, and replied, "It sure is."

EPILOGUE

—◆—

And the Beat Goes On

*It has been said, "time heals all wounds." I do
not agree. The wounds remain. In time, the mind,
protecting its sanity, covers them with scar tissue and
the pain lessens. But it is never gone.*

—ROSE FITZGERALD KENNEDY

*I*t is sometimes difficult to comprehend that I am the only member of our original family still living. Anyone who has lost someone they love understands the feeling. The spirit and laughter and intellect of the people you loved were so alive and so present that it is impossible to realize that they will not grace the earth and your life again.

My parents and brothers and sisters had the earliest and most profound influence on my life, and I remain so proud of what they accomplished in the years that followed those special, yet too-short days that we all spent together. Joe's legacy as a war hero is forever etched in history. Mother and Dad always cautioned against our imagining what he would have become if he had lived, yet I have no doubt his impact would have been profound.

The same is true of Kathleen. Kick returned home with us, but she wanted to stay in England. She had become very fond of a young British soldier, William "Billy" Cavendish, the Marquess of Hartington, and she was anxious to see him again, as well as her many friends. Mother and Dad eventually agreed to allow Kick to return to London, at the height of the war, to work in the Red Cross Club.

The romance between Kick and Billy was the source of tremendous chatter among the English social set. Kick was a Catholic and Billy was in line to be the next Duke of Devonshire, a critical position in the Church of England. At that time in history, the prospect of such a pairing raised eyebrows and attention.

"Everyone in London is buzzing with rumors, and no matter what happens we've given them something to talk about," Kick reported in a letter to Jack.

Mother recalled how concerned she and Dad were about the

Kick in her Red Cross uniform (1944)

situation, as were Billy's parents, the Duke and Duchess of Devon-shire. They called and wrote and sent cables to each other over the matter. They were good friends and they wanted the best for their children. "The point of it all was simply ... to find a way through innumerable complications to make possible a marriage" for two young people who were "much in love," Mother recalled.

Rosemary and Dad (London, 1938)

I was studying at the Sacred Heart Convent in Noroton, Connecticut, at the time and, of course, caught wind of the concern over
Kick. Dad telephoned the school and asked if he could come see me
during visitors' day on Sunday. I was thrilled as always to have
him to myself and, after lunch, he asked if I would walk down
with him across the lawn toward Long Island Sound. There, as

we strolled along the rocky shore, Dad told me that Kick and Billy had decided to get married. They would not marry in the Catholic Church but rather would have a civil ceremony in London.

I did not know how to take the news. The Church's teaching to marry someone of our own faith was engrained in all of us, yet I adored my older sister. Did this mean I would have to let her go? Could she still go to heaven?

Dad immediately put me at ease:

"God loves Kick and she's a great girl," he said. "Say a prayer for her and be happy for her.

"It will all be okay."

$\mathcal{M}$y brother Joe, a lieutenant in the U.S. Navy, was posted in London and gave Kick away at her wedding. He called us before the ceremony for advice on what to wear. Then immediately afterward he called again to report on the day's events. We gathered around as Mother, ear pressed to the phone, repeated Joe's account of the ceremony:

"Joe says Kick looked absolutely lovely and carried a bouquet of pink camellias. She was floating on air!"

We were all thrilled for her.

—◆—

$\mathcal{I}$n the end, Kick and Billy's time together was painfully short. They spent only one month as newlyweds before Billy had to return to combat in Europe. He was killed in action, just one month after we lost Joe.

Kick was destroyed to lose her beloved husband, as well as her brother. She came to stay with our family in Hyannis Port for a short time, but soon decided to go back to London, where she felt closest to Billy.

Kick made a life for herself there among Billy's family and her many friends. She wrote to us often and various members of the family had a chance to visit her over the three years that followed. It is so hard to fathom that we then lost her too. Kick was on her way to meet Dad in France when her plane went down. It was the ultimate heartbreak. Learning of her death, our shattered dad wrote of his sparkling daughter: "We know so little about the next world that we must think that they wanted just such a wonderful girl for themselves."

Little could we understand as well the sadness that befell our beloved Rosemary. She had been doing so well in her school in England, and we had such hope for her continued improvement. However, even after she returned home, her anxieties continued to increase. It greatly worried Dad and Mother, who both loved her so. Dad began an intense period of research, seeking out the best doctors to help Rosemary.

Dad read about a new procedure called a lobotomy and, after speaking with multiple professionals about it, became confident it would help ease his daughter's growing agitation. To minimize risk, Dad went straight to the acclaimed expert, Dr. James Watts, to perform the surgery. It went tragically wrong. It is still not clear what happened. Rather than finding relief through the procedure, Rosemary lost most of her ability to walk and communicate. We had been so hopeful, and were devastated.*

---

*   In 1935, the Portuguese neurologist António Egas Moniz performed a new procedure called a leucotomy, which was intended to treat mental illness. The American psychiatrist and physician Walter J. Freeman, subsequently adapted the procedure, renaming it the lobotomy.

Freeman's partner, neurosurgeon James Watts, performed the first lobotomy in 1936. The pair reported significant improvement in patients' conditions and dispositions, leading medical experts across the nation to laud it as a breakthrough procedure. Moniz would go on to win the Nobel Prize in Physiology or Medicine because of his work. Time would tell, however, that the lobotomy was not the breakthrough it was believed to be. We later learned of other patients who had suffered the same terrible fate as Rosemary, including Rose Williams, the sister of the playwright Tennessee Williams.

It is easy from this vantage point to second-guess decisions, but in that time and place our parents made theirs with advice from the leading medical professionals of the day. As a family, we could only do one thing in the aftermath: move forward by finding the best care available for Rosemary—the wonderful sisters of St. Coletta at the School for Exceptional Children in Wisconsin—and by finding ways to help other people with disabilities. Dad and Mother founded the Joseph P. Kennedy Jr. Foundation in memory of Joe, and dedicated it to providing leadership and encouraging research in the field of disability. All of us siblings and our children visited Rosemary at her home in Wisconsin, and she would come east to visit us on holidays, spending Eastertime at our house with my family. As always, Dad was at the forefront of our unified work for and in the name of Rosemary, but he remained heartbroken over the outcome of her surgery for the rest of his life.

These early tragedies were devastating for all of us. Yet encouraged by our parents' faith and example, we continued on. As it does, time eventually brought other, happier moments, as well as fun and accomplishments. Jack rose to the highest office in the land when he was elected president in 1960. He reinvigorated a tired nation and ignited it to service.

Jack's death was intensely personal for each of us. Yet he also belonged to our country, for which his death was intensely personal as well, especially among the young. As Mother later wrote, "So many young people looked to Jack as a hero and model for their lives. We in the family reacted to our common grief in our own ways." She recalled:

> *But we could all be reasonably steady because of the faith, hope, and love we shared. And because we knew quite well what Jack would want from us: He would want courage, he would want as many smiles as we could manage, and he would want his death to be an affirmation of life. He would want us to think of him with love, but to live for the living, and to cherish such happiness as we could find and give, and any bit of light and hope and example of fortitude that might help us and others. We all realized this, and it needed no discussion, and we all did our best.*

Bobby, the one everyone turned to for support, was Jack's right-hand man and confidant. At Jack's request, Bobby stepped up to serve as attorney general after his election. Following Jack's death, and a particularly penetrating period of grief, Bobby decided to run for office himself. In 1965 he was elected U.S. senator

Painting by Teddy of his sailboat, the *Mya*

for the state of New York, and three years later, in the throes of a nation in strife and at war, he launched a bid for the presidency.

Bobby's love of his fellow man was formed in his youth. His was a courageous voice on civil rights and for the underprivileged of this country. His was also a life taken too, too soon.

His friend and adviser Richard Neustadt wrote of riding on

Bobby's funeral train from New York to Washington, DC, and seeing the thousands of people along the tracks paying their respects, "silent, serious, absorbed and concerned people."

They were, "by all appearances, by every sign of face and dress, primarily the urban poor, predominately the ghetto poor. Surely more than half—by some proportion I don't know—more than half those faces were black.

"He lived by the belief that individuals can make a difference, that individual will can be imposed on situations, that a man can matter if he mobilizes all his effort, focuses his mind and stamina and influence, holds nothing back," Neustadt wrote. "Bob had no naïve confidence that effort on his part could guarantee success, but I think he had enormous confidence that effort, focused effort, could impose upon the world some share of the conditions making for success—and thus could move us forward—and those crowds along the railroad vindicate his confidence.

"No man has lived his life in vain if he can leave that testimony for the rest of us."

We grieved again in private, but also with a nation. Then, in Bobby's spirit, we again carried on.

Eunice, that indomitable force, inspired by our sister Rosemary, devoted her life to helping people with intellectual challenges fully integrate into society. She was very active with the Joseph P. Kennedy Jr. Foundation and started holding sports camps at her home in Maryland that quickly became a model for others throughout the nation. Just one month after Bobby's death in 1968, Eunice announced the formation of the Special Olympics, the celebrated organization that now gives millions of people with disabilities in 170 nations around the world the opportunity to participate in sports. Today, it and the Foundation are led by Eunice's son Tim, who, with the support of his brothers, sister, and cousins, does a fantastic job ensuring access and respect for all people with disabilities.

Pat, my protector, cohort, and friend, whose infectious wit and grace matched her natural glamour, turned her early love for the arts into a vocation, helping support struggling playwrights and founding the National Committee for the Literary Arts to provide lectures and scholarships.

And Teddy. Our little brother with heart, energy, and intellect to spare, joined his brothers in the U.S. Senate, where he remained for nearly forty-seven years, representing the poor, disadvantaged, and forgotten both in his home state of Massachusetts and across the nation and world.* Rather than crumbling under the loss of

---

* Teddy was first elected in 1964 and reelected seven more times. He was the fourth-longest serving senator in U.S. history at the time that he died in 2009.

Our family (1939)

Jack and Bobby, he found the strength to support all of us who were left behind.

Teddy was a devoted father to his own three children, and to Jack's and Bobby's as well. He felt acutely the hole that was left in their lives by the deaths of their fathers. From his tremendous well of compassion, Teddy took on the responsibility of watching out for them and bringing us together for companionship and fun.

Each year, Teddy organized elaborately involved trips for all of us to America's historic sites, renting several buses and arranging all the meals, accommodations, and tours. At the time, we simply thought this was Teddy doing what Teddy does, leading the way and making sure all the children felt the connection to their family. But in retrospect, I find it incredible how he managed to do all this in the midst of the many critical issues of national importance he was dealing with daily at the Senate. He not only always found time for us, he made us feel as if there was nothing else he would rather be doing.

With Teddy at the lead, dozens of us, grown-ups and children, spent summers tromping around Walden Pond in Concord, Massachusetts, through the Norman Rockwell Museum in Stockbridge, across the battlefields at Appomattox and Gettysburg, over the Brooklyn Bridge, and to the George Bush Presidential Library in Texas, where President George and First Lady Barbara Bush welcomed us so graciously. Teddy even took us on a hysterical canoe-ing and camping trip that had no end of mishaps.

Following in Mother's footsteps, Teddy would read up on all the historical facts of each site prior to striking out on a new ad-venture. Sometimes he would invite a historian along to guide us, but at other times, Teddy would narrate the stories himself. Young nieces and nephews hung on his words, carried afloat by his broad knowledge, enthusiasm, and wit.

"And along these streets, in the dark of night, Paul Revere mounted his horse," Teddy would whisper. Then he burst into a roar that made all of them jump: "The British are coming! The British are coming! The British are coming!"

The story passed down to another generation.

Joe, Jack, Rosemary, Kathleen, Eunice, Pat, Bobby, and Teddy—all are now gone, some taken in the prime of their lives, others in quiet peace in later years. We lost Dad, our giant and our leader, in 1969, following a series of strokes. Mother, ever forceful and determined, remained with us many years longer, until the age of 104, when she died peacefully during a rest at her beloved Hyannis Port home.

*F*or some time now, when I travel to Washington, DC, I stop on my way to and from the airport at Arlington National Cemetery to pay my respects to my four brothers. All of them— Joe, Jack, Bobby, and Teddy—are memorialized together at Arlington, their markers a testament to the tremendous gifts they gave to their country. The generations that come after them are called to do the same.

Mother and Dad had twenty-eight grandchildren in all. These grandchildren, and their children and grandchildren, live very different lives from their forebears, who rode to school on horses, pinched pennies for phone calls, and traveled hours by train. Yet the problems that they and the world face today are no different. These problems carry the same burden and offer the same opportunity. They require the same determination, focus, and drive that they required of my grandparents, parents, brothers, and sisters, and all the young people of generations before.

More than seventy-five years ago, my father, far away in England, wrote to my brother Bobby, a young boy of fourteen, at the height of World War II:

"It is boys of your age who are going to find themselves in a very changed world, and the only way you can hold up your end is

to prepare your mind so that you will be able to accept each situation as it comes along. So don't, I beg of you, waste any time. Do all
the things necessary to get yourself in good physical condition—
and work hard."

Amen.

ACKNOWLEDGMENTS

—◆—

*T*his book was born of an idea to capture on paper a period of time that was so important to my life and to the life of my family. It was inspired by the example of my parents who loved each other dearly and guided the nine of us through life's joys and sorrows with forbearance, wisdom, and faith. Mother and Dad taught us to be thankful to those who came before us and to give back to our fellow man and country. They taught us to never take anyone or anything for granted. I wanted to remember them, and my brothers and sisters, as they really were, and I am grateful to all those who helped make that possible.

First and foremost, I want to thank my friend and partner in this adventure, Amy Seigenthaler, who helped me shape these stories into the collection they became. Our laughs kept us going, even on those days when we were close to hanging it all up. What fun we had.

And to my agent, Laurie Abkemeier, my editor, Gail Winston,

and everyone at Harper, especially Sofia Groopman—I am extremely grateful for the confidence they showed in the material and the care they took shepherding the book to completion.

Thank you also to the librarians and archivists who helped source images for the book, especially Connor Anderson, Laurie Austin, and Maryrose Grossman at the John F. Kennedy Presidential Library and Museum; Jennifer Quan at the John F. Kennedy Library Foundation; and Kirsten Carter at the Franklin D. Roosevelt Presidential Library and Museum—a job well done.

Proceeds from the sale of the book will be donated to the restoration of the USS *Joseph P. Kennedy Jr.*, the ship I christened in 1945 in memory of my oldest brother and godfather, Joe. The ship is now in need of repair and, on behalf of my entire family, I would like to contribute to that effort. I know this would make our parents, and all the brave men who served on her, very happy.

While writing this book, it has been enlightening for me to look back at my childhood and realize how much has changed in my lifetime. I am grateful to my mother for urging us to keep diaries and write letters when we were young, and I encourage parents today to do the same with their children. Suggest that they write about their adventures, their feelings, and the sights around them. It will be remarkable for them to look back in twenty, thirty, or fifty years and see how different the world is. I also encourage grandparents to share their stories of growing up. Years from now, their grandchildren will still hold fast to those stories as reminders of the people and the past that helped make them who they became.

IMAGES AND CREDITS

—◆—

Painting by Ted Kennedy of the Kennedy home in Hyannis Port, frontis: Painting by Ted Kennedy. David Mager Photography.

Joe, Jack, Rosemary, Kick, Eunice, Pat, Bobby, and me, page 4: Richard Sears. Pathe News Boston. Kennedy Family Collection. John F. Kennedy Presidential Library and Museum, Boston.

Mother and Dad, page 11: The John F. Kennedy Presidential Library and Museum, Boston.

The Nine of Us, with Mother and Dad, page 16: Bachrach/Getty Images

Joe, page 19: John F. Kennedy Library Foundation. Kennedy Family Collection. John F. Kennedy Presidential Library and Museum, Boston.

Jack, page 21: The John F. Kennedy Presidential Library and Museum, Boston.

Rosemary, page 23: Angus McBean Photograph. © Houghton Library, Harvard University.

Kick, page 25: Sport & General Press Agency, Ltd. Kennedy Family Collection. John F. Kennedy Presidential Library and Museum, Boston.

Eunice, page 27: John F. Kennedy Library Foundation. Kennedy Family Collection. John F. Kennedy Presidential Library and Museum, Boston.

Pat, page 28: John F. Kennedy Library Foundation. Kennedy Family Collection. John F. Kennedy Presidential Library and Museum, Boston.

Bobby, age 30: John F. Kennedy Library Foundation. Kennedy Family Collection. John F. Kennedy Presidential Library and Museum, Boston.

Mother holding me, with Bobby, page 32: Photographer Unknown. Kennedy Family Collection. John F. Kennedy Presidential Library and Museum, Boston. David Mager Photography.

Teddy and Bobby, page 33: John F. Kennedy Library Foundation. Kennedy Family Collection. John F. Kennedy Presidential Library and Museum, Boston.

Eunice, Joe, Jack holding me, Rosemary, Bobby, and Pat, page 35: John F. Kennedy Library Foundation. Kennedy Family Collection. John F. Kennedy Presidential Library and Museum, Boston.

Grandpa John "Honey Fitz" Fitzgerald, page 38: Photographer Unknown. Kennedy Family Collection. John F. Kennedy Presidential Library and Museum, Boston

P. J. Kennedy, page 41: Photographer Unknown. Kennedy Family Collection. John F. Kennedy Presidential Library and Museum, Boston.

Dad, page 42: Photographer Unknown. Kennedy Family Collection. John F. Kennedy Presidential Library and Museum, Boston.

Pat and Teddy, page 44: John F. Kennedy Library Foundation. Kennedy Family Collection. John F. Kennedy Presidential Library and Museum, Boston.

Teddy, Bobby, and me, page 45: John F. Kennedy Library Foundation. Kennedy Family Collection. John F. Kennedy Presidential Library and Museum, Boston.

Teddy and me, page 61: John F. Kennedy Library Foundation. Kennedy Family Collection. John F. Kennedy Presidential Library and Museum, Boston.

Dad, page 74: © Hulton-Deutsch Collection/CORBIS.

Jack, me, Mother, Dad, Pat, Bobby, Eunice, and Teddy, page 81: The John F. Kennedy Presidential Library and Museum, Boston.

Grandma, Grandpa, and Mother, page 87: Richard Sears. Kennedy Family Collection. John F. Kennedy Presidential Library and Museum, Boston.

Bobby and me, page 101: Richard Sears. Kennedy Family Collection. John F. Kennedy Presidential Library and Museum, Boston.

Teddy and Jack, page 104: John F. Kennedy Presidential Library and Museum, Boston.

Bobby and Mother, page 110: Illustrated London News, Ltd. Kennedy Family Collection. John F. Kennedy Presidential Library and Museum, Boston.

Bobby, page 112: John F. Kennedy Library Foundation. Kennedy Family Collection. John F. Kennedy Presidential Library and Museum, Boston.

Bobby, page 113: John F. Kennedy Library Foundation. Kennedy Family Collection. John F. Kennedy Presidential Library and Museum, Boston.

Letters between President Roosevelt and Bobby, 119–121: Courtesy of the Franklin D. Roosevelt Library.

Bobby, Mother, Pat, page 123: John F. Kennedy Library Foundation. Kennedy Family Collection. John F. Kennedy Presidential Library and Museum, Boston.

Eunice, Dad, Kick, and Pat, page 134: Bert Morgan/Getty Images.

Kick, page 138: Photographer Unknown. Kennedy Family Collection. John F. Kennedy Presidential Library and Museum, Boston.

Pat, page 141: The John F. Kennedy Presidential Library and Museum, Boston.

Me in Scottish Dress, page 147: John F. Kennedy Library Foundation. Kennedy Family Collection. John F. Kennedy Presidential Library and Museum, Boston.

Mother and Eunice, page 148: Bert Morgan/Getty Images.

Mother and Teddy, page 154: Photographer Unknown. Kennedy Family Collection. John F. Kennedy Presidential Library and Museum, Boston.

Mother, Teddy, and me, page 164: AP Images.

Mother, page 171: Frank Turgeon, Jr., Palm Beach, FL. Kennedy Family Collection. John F. Kennedy Presidential Library and Museum, Boston.

Joe, Dad, and Jack, page 177: The John F. Kennedy Presidential Library and Museum, Boston.

Jamie Wyeth painting of Teddy on his sailboat, Mya, page 189: © Jamie Wyeth. David Mager Photography.

Mother and Dad, page 194: John F. Kennedy Library Foundation. Kennedy Family Collection. John F. Kennedy Presidential Library and Museum, Boston.

Kick, Dad, Mother, Pat, me, Bobby, and Teddy, page 206: © Daily Mirror/Mirrorpix/Corbis.

Teddy and me, page 207: AP Images.

Kick, Rosemary, and Mother, page 209: © CORBIS.

Our family at the Vatican, page 210: © Bettmann/CORBIS.

Joe, page 214: Photographer Unknown. Kennedy Family Collection. John F. Kennedy Presidential Library and Museum, Boston.

Me christening the USS Joseph P. Kennedy, Jr., page 218: © Bettmann/CORBIS.

Ethel and Bobby, page 219: Miller Studio, Quincy, MA.

Steve and me, page 220: Beulah Harris Ingall Collection.

Stephen Jr., Amanda, Steve, me, Kym, and William, page 221: Frank Teti, Teti-Miller Collection. John F. Kennedy Presidential Library and Museum, Boston.

Campaigning for Jack, page 239: © CORBIS.

Our family the day Jack won the election, page 231: Popperfoto/Getty Images.

Jack and Steve, page 232: Robert Knudsen. White House Photographs. John F. Kennedy Presidential Library and Museum, Boston.

Irish President Sean Lemass, Mrs. Kathleen Lemass, me and Jack, page 234: White House Photographs. David Mager Photography.

Kick, page 239: The John F. Kennedy Presidential Library and Museum, Boston.

Rosemary and Dad, page 240: © Bettmann/CORBIS.

Painting by Teddy of his sailboat, page 246: Painting by Ted Kennedy. David Mager Photography.

Our family, page 249: Photographer Unknown. Kennedy Family Collection. John F. Kennedy Presidential Library and Museum, Boston.

ABOUT THE AUTHOR

—◆—

JEAN ANN KENNEDY SMITH is the former United States Ambassador to Ireland and founder of VSA, an international organization that provides arts and education opportunities for people with disabilities and increases access to the arts for all. The eighth of nine children born to Joseph P. Kennedy Sr. and Rose Fitzgerald Kennedy, she is a mother of four and widow of the late Stephen Smith. She lives in New York.